AF849308

Implementing a Prototype Energy and Water Quality Management System

Prepared by:
Lawrence A. Jentgen, **P.E.** and **Steve Conrad**
EMA, Inc., Tucson, AZ 85718
Richard Riddle
EMA, Inc., St. Paul, MN 55113
Eric Von Sacken and **Kelvin Stone**
Colorado Springs Utilities, Colorado Springs, CO 80903
Walter Grayman, P.E., Ph.D.
W.E. Grayman Consulting Engineer, Cincinnati, OH 45229
and
Satish Ranade, Ph.D.
Las Cruces, NM 88001

Jointly sponsored by:
Awwa Research Foundation
6666 West Quincy Avenue
Denver, CO 80235-3098
and
Colorado Springs Utilities
2855 Mesa Road
Colorado Springs, CO 80904-1199

Published by:

Awwa Research Foundation
Advancing the Science of Water®

IWA Publishing

DISCLAIMER

This study was funded by the Awwa Research Foundation (AwwaRF). AwwaRF assumes no responsibility for the content of the research study reported in this publication or for the opinions or statements of fact expressed in the report. The mention of trade names for commercial products does not represent or imply the approval or endorsement of AwwaRF. This report is presented solely for informational purposes.

Published by IWA Publishing, Alliance House, 12 Caxton Street, London SW1H 0QS, UK

Telephone: +44 (0) 20 7654 5500; Fax: +44 (0) 20 7654 5555; Email: publications@iwap.co.uk;
Web: **www.iwapublishing.com**

AwwaRF report number 90948F. Originally published by AwwaRF for its subscribers in 2003.
IWA Publishing version published 2004.

Copyright © 2003
by
Awwa Research Foundation

ISBN 1 84339 857 5

Implementing a Prototype Energy and Water Quality Management System

The mission of the Awwa Research Foundation (AwwaRF) is to advance the science of water to improve the quality of life. Funded primarily through annual subscription payments from over 1,000 utilities, consulting firms, and manufacturers in North America and abroad, AwwaRF sponsors research on all aspects of drinking water, including supply and resources, treatment, monitoring and analysis, distribution, management, and health effects.

From its headquarters in Denver, Colorado, the AwwaRF staff directs and supports the efforts of over 500 volunteers, who are the heart of the research program. These volunteers, serving on various boards and committees, use their expertise to select and monitor research studies to benefit the entire drinking water community.

Research findings are disseminated through a number of technology transfer activities, including research reports, conferences, videotape summaries, and periodicals.

CONTENTS

TABLES

FIGURES

FOREWORD

The Awwa Research Foundation is a nonprofit corporation that is dedicated to the implementation of a research effort to help utilities respond to regulator requirements and traditional high-priority concerns of the industry. The research agenda is developed through a process of consultation with subscribers and drinking water professionals. Under the umbrella of a Strategic Research Plan, the Research Advisory Council prioritizes the suggested projects based upon current and future needs, applicability, and past work; the recommendations are forwarded to the Board of Trustees for final selection. The foundation also sponsors research projects through the unsolicited proposal process; the Collaborative Research, Research Applications, and Tailored Collaboration programs; and various joint research efforts with organizations such as the U.S. Environmental Protection Agency, the U.S. Bureau of Reclamation, and the Association of California Water Agencies.

This publication is a result of one of these sponsored studies, and it is hoped that its findings will be applied in communities throughout the world. The following report serves not only as a means of communicating the results of the water industry's centralized research program but also as a tool to enlist the further support of the nonmember utilities and individuals.

Projects are managed closely from their inception to the final report by the foundation's staff and large cadre of volunteers who willingly contribute their time and expertise. The foundation serves a planning and management function and awards contracts to other institutions such as water utilities, universities, and engineering firms. The funding for this research effort comes primarily from the Subscription Program, through which water utilities subscribe to the research program and make an annual payment proportionate to the volume of water they deliver and consultants and manufacturers subscribe based on their annual billings. The program offers a cost-effective and fair method for funding research in the public interest.

A broad spectrum of water supply issues is addressed by the foundation's research agenda: resources, treatment and operations, distribution and storage, water quality and analysis, toxicology, economics, and management. The ultimate purpose of the coordinated effort is to assist water suppliers to provide the highest possible quality water economically and reliably. The true benefits are realized when the results are implemented at the utility level. The foundation's trustees are pleased to offer this publication as a contribution toward that end.

This project was jointly funded by the Awwa Research Foundation (AwwaRF) and Colorado Springs Utilities (CSU). AwwaRF is a non-profit corporation that is dedicated to the implementation of research to help utilities respond to regulatory requirements and high-priority concerns of the industry. CSU is a municipally owned corporation that provides water, wastewater, gas, and electric service to the citizens of Colorado Springs, Colorado.

As an industry, water utilities collectively represent the largest single user of electricity in the country, using an estimated 75 billion kWh annually, about three percent of all electric power generated in the United States. Yet the typical water utility has been a relatively unsophisticated user of energy, devoting little attention to strategies aimed at reducing energy costs. Recent energy crises have caused some utility managers to open their eyes to the importance of energy management. These crises have coincided with the need for many utilities to expend major capital appropriations on infrastructure due to growth in their service areas and to rehabilitate aging facilities. This two-pronged attack on utilities' economic resources has the water industry seeking answers. One possible strategy is to implement an Energy and Water Quality Management System (EWQMS).

An EWQMS is defined as a collection of individual software applications and operational processes that solve water quality and energy problems encountered in the effort to efficiently operate a water supply system. An EWQMS facilitates alignment of a utility's operational functions by applying structure to planning and scheduling activities by means of a formal Daily Operating Plan, which is available to and followed by all operations units in the utility. Operational "silos" are invalidated and redundant or ad hoc work activities are eliminated though system planning and scheduling. Utilities that implement an EWQMS can expect lower energy costs, reduced operations and maintenance budgets, enhanced water quality, greater availability of capital for system improvements, and a heightened reputation for customer responsiveness.

This report defines the experience of Colorado Springs Utilities (CSU) in the implementation of an EWQMS, to serve as a prototype for other water utilities. Readers of this report will learn how CSU justified an EWQMS from an energy and water quality/supply perspective. The approach used to define the operational process changes necessary to enable an EWQMS are also presented. EWQMS software specifications are included along with a description of the methods used to procure, develop, and install the software.

This publication provides a "head start" for any water utility contemplating an EWQMS. The concept of an integrated EWQMS is new to the water utility industry, and the approach and lessons learned by the authors can reduce the time and cost to implement organizational changes and technology to realize improved operational effectiveness.

Edmund G. Archuleta, P.E.
Chair, Board of Trustees
Awwa Research Foundation

James F. Manwaring, P.E.
Executive Director
Awwa Research Foundation

ACKNOWLEDGMENTS

The authors of this report are indebted to CSU management who supported the EWQMS program: Leah Ash, Bob Robler, Scott Campbell, Cathy Lee, and Curtis Mitchell. The CSU EWQMS Implementation Team played a critical part in definition and implementation of the organizational change required to enable the EWQMS. The following CSU staff participated on the EWQMS Implementation Team: Jean Mueller, Kelly Stone, Lisa Barbato, Tick Benz, Donna Bellows-Parks, Jim Browne, Kevin Hershey, Chuck Johnson, Mike Palmer, John Rae, and Russ Brailsford.

The authors also acknowledge the advice and support from the Project Advisory Committee (PAC):

Randy Hawley, Portland Water Bureau, Portland, Ore.
Bob Sullivan, Las Vegas Valley Water District, Las Vegas, Nev.
Miriam Heller, National Science Foundation, Washington, D.C.
Joan Kersnar, Seattle Public Utilities, Seattle, Wash.

Last, but not least, the authors wish to acknowledge Lisa Moher who facilitated workshops and prepared the EWQMS Use Cases, and Laura Botkin, Linden Hickey, and Laura Riegel who edited and prepared the manuscript.

Lawrence Jentgen, P.E.	Eric Von Sacken	Emerson Lomaquahu
Principal Investigator	Project Manager	Project Manager
EMA, Inc.	Colorado Springs Utilities	Awwa Research Foundation

EXECUTIVE SUMMARY

The operating environment for water utilities is changing dramatically in the new century. Competition motivates operations and maintenance managers to continually pursue improved methods for reduction of operating costs. Electric and gas deregulation present both a risk and competitive opportunity for energy cost reduction for those water utilities that can more optimally manage and control their systems.

Water quality and supply issues must also be considered when working to minimize operational costs. The performance criteria defined by "quality" and "supply" are the boundaries within which optimization problems can be initiated and solved.

New information and control systems and operating procedures are required to manage water utility operations in this changing environment. Based on these requirements, the notion of an Energy and Water Quality Management System (EWQMS) was created in three previous projects jointly sponsored by AwwaRF and the Electric Power Research Institute (EPRI). The first project defined the needs, requirements, and potential opportunities for an EWQMS. The second project, at East Bay Municipal Utility District (EBMUD), determined the economic potential and defined a model for an EWQMS. The third project developed EWQMS requirements and functional specifications.

In this project, an implementation prototype for EWQMS, was developed at Colorado Springs Utilities (CSU). It is the logical next step in the EWQMS program progression.

RESEARCH OBJECTIVES

The approach to the CSU EWQMS project is built around the concept of the Operations Planner Scheduler (OPS). The OPS is one or several employees, supported by software, who develop and execute an optimum daily operating plan for the water system. The optimum daily operating plan is created manually at first, using simple office automation tools, then expanded using more complex software. The actual operation of the OPS was modeled at CSU using the object-oriented concepts developed in the AwwaRF Utility Business Architecture (UBA) Project.

Building the EWQMS on the foundation of the OPS function provides bridges between technology, people, and the business of operating CSU. The OPS also provides the means to implement the human infrastructure necessary to construct the EWQMS technical infrastructure. CSU has implemented an Advanced Integrated Supervisory Control and Data Acquisition System (AI SCADA) in parallel with the EWQMS, which provides the enabling technology to integrate EWQMS concepts into daily operations.

Another important aspect of the EWQMS implementation at CSU is water quality and supply. CSU has significant operating challenges associated with water quality and water supply conditions. The OPS models must be developed with an emphasis on daily water quality management.

A significant opportunity to advance the EWQMS for the water utility industry is the close relationship between the electric and water operational functions at CSU. Working with operations personnel from within a single organization promised to minimize barriers to collaboration that can result from the more usual supplier/customer relationship between water and electric utilities. The discoveries made can then be extended to apply to the "normal" situations found in the industry at large. Significant opportunities have been identified with control of hydroelectric generation and distribution system pumping as a strategic energy management initiative. Immediate benefits are projected for both water and electric in the current business environment of energy deregulation and competition. Based on this background, five project objectives were defined:

1. Quantify the projected benefits of an EWQMS using the methodology defined in previous AwwaRF/EPRI reports.
2. Design, model, implement, measure results, and document the OPS function previously modeled in the EWQMS project.
3. Evaluate AwwaRF-generated EWQMS specifications, procure and implement software systems as defined in the EWQMS project. Systems implemented will be those that can be operational within a short period of time and show a positive return on investment.
4. Achieve significant operating cost savings and improved performance for the water utility industry and implement methodologies and systems that yield a positive return on investment.
5. Document the process of EWQMS development in a Final Report and release AwwaRF specifications used for the project.

Anticipated results of this project included significant energy cost reduction and better management of water quality and supply associated with actual demonstration of the EWQMS. Lessons learned in the course of implementing the OPS and other components of the EWQMS at CSU will apply industry-wide and accelerate development of the EWQMS concept for the water utility industry.

APPROACH

The first step in the project was to orient CSU management and the PAC on the background of and opportunities related to an EWQMS in order to establish a common reference for this new concept of water operations. Next, the project team analyzed water system SCADA data, energy billing information, electric hydroelectric generation production records, and water quality data to quantify the expected benefits of an EWQMS. Energy and water quality modeling was performed to assess operating scenarios to improve water quality and optimize the use of energy at CSU. The results of these analyses were prepared in report and presentation format for CSU management.

Several meetings were held with CSU upper management to present the potential opportunities and the estimated cost of an EWQMS. After close scrutiny, CSU management authorized the creation of an EWQMS Implementation Team (I-Team) as part of a water utility-wide optimization project.

The EWQMS I-Team then defined the operational processes or use-cases necessary to implement the software and organizational changes for an EWQMS. This modeling process and previously completed benefits analysis provided the necessary information to develop functional specifications from the software applications.

After development of the functional specifications, work commenced on the EWQMS software applications. The following applications were developed:

- Consumption Forecaster
- Economic Pump/Valve Controller
- Water System Simulator
- Optimum Pump Scheduler
- Water Quality Analyzer
- Water Source Analyzer
- Daily Operating Plan
- Daily Operating Plan NetConference
- Construction/Maintenance Clearance Process

Operational changes were also implemented to facilitate the concept of an OPS. The first process placed into operation was the clearance request procedure.

Experience gained from the implementation of the EWQMS at CSU was documented and included in this report.

CONCLUSION

The following are conclusions of this EWQMS Implementation Prototype at CSU:

1. The tangible benefits are significant. The annual opportunity for hydroelectric generation and pump/valve optimization is an estimated $550,000 per year. The escalating cost of energy during this period should result in even greater savings.
2. The return on investment for software is less than three years.
3. Implementation of the Water Supply Analyzer will substantially improve the daily management (cost, water rights, quality) of the multiple surface water sources supplying the CSU water system.
4. Better management of the water sources and distribution storage will result in lower treatment costs.
5. Implementation of the formal clearance procedure associated with operations planning and scheduling will improve system reliability, enhance the safety of maintenance crews, and facilitate a "systems" perspective for program driven maintenance.
6. The single point of operational coordination provided by the OPS function consolidates the water system security process.
7. Moving the CSU organization from independent to coordinated interdependent operations management is a significant cultural change requiring patience, understanding, and time.
8. The operating environment of the future is projected to be more dynamic and complex requiring more coordinated, better planned, and agile operations to minimize cost and provide high quality service to water customers.

RECOMMENDATIONS

The following are recommendations for a water utility considering an EWQMS:

1. Assess the opportunity and quantify the tangible and intangible benefits of an EWQMS. The benefits provide the business case for EWQMS implementation.
2. Provide a clear and concise report to management regarding the opportunities and cost of an EWQMS.
3. Define a vision for future operations that includes an EWQMS.
4. Develop an EWQMS model based on the AwwaRF reference model.
5. Define and model the energy and water quality/supply management processes associated with operations planning and scheduling.
6. Implement the organizational process change first, then the EWQMS software technology.
7. Develop functional software specifications based on the benefits analysis and operational modeling.
8. Implement software solutions incrementally. Start first with more simplistic solutions and build out with more complexity if there is a commensurate return on investment.
9. Accurate SCADA data is required for operational models, and resources must be applied to install and calibrate field instrumentation.
10. Time to tune and calibrate EWQMS models is a function of water system complexity and the availability and accuracy of SCADA data.
11. Obtain management sponsorship and support to ensure project success.

FUTURE RESEARCH

This project concluded with the development and installation of EWQMS software and development of organizational processes. As with most prototypes, this one will require further tuning, including operational testing, analysis of results, and refinements to both the software and organizational processes to realize the vision and projected benefits.

It is recommended the next EWQMS project develop and document the processes required to place an organization and software in full operational mode. The water utility industry has little experience with operations optimization systems, and experience gained in a future research project would provide high value to utilities considering and implementing an EWQMS.

CHAPTER 1
INTRODUCTION AND BACKGROUND

INTRODUCTION

This chapter provides an introduction to the Energy and Water Quality Management System (EWQMS) at Colorado Springs Utilities (CSU). A broad understanding of EWQMS concepts is required prior to solving specific problems associated with energy and water quality optimization for the water system. Knowledge of the interrelationships between the water system, the people who operate the system, and the software that will be used to plan and schedule utility operations is important to meet the specific needs of the utility.

An overview of the generic concept of an EWQMS is first presented in this chapter, followed by a description of the Colorado Springs Water System, overview of the CSU EWQMS, objectives of the CSU EWQMS, and organization of the report.

OVERVIEW OF AN EWQMS

The Electric Power Research Institute (EPRI) and Awwa Research Foundation (AwwaRF) introduced the concept of an EWQMS in a February 1997 report titled, "Energy and Water Quality Management System Collaborative Project for Water System Optimization." This report defined the functionality of an EWQMS.

An EWQMS is a collection of individual application software programs. Together these programs provide information used by utility operations staff to solve water quality, supply and energy management problems associated with carrying out the utility's mission. EWQMS users receive information and prepare plans for daily decision-making. These plans are developed using optimization and simulation techniques embedded in the software programs. These techniques augment the knowledge of individuals in operations, planning, water quality, etc.

Conceptually, these software programs are closely coupled to the Supervisory Control and Data Acquisition (SCADA) system. While they require data from the SCADA system, these programs are typically used by individuals outside of a utility control room (e.g., operations planning, water quality engineering, energy use monitoring, and planning). Their principal use is to identify the least cost operating method, from the perspective of the entire water system, while still complying with water quality and supply constraints.

Figure 1.1 is a functional overview of an EWQMS presented in an Electric Power Research Institute (EPRI) report, *Energy and Water Quality Management System: Collaborative Project for Water System Optimization* (Jentgen et al. 1997). This generic model may be used by the water utility considering an EWQMS. It consists of the following functional components and their associated data flows:

Starting at the left part of Figure 1.1, the Operations Planner and Scheduler (OPS) develops the daily operating plan for the entire water system. This function prepares the production schedule for treatment plants and pumping plant schedules. Each day an operating plan is given to the treatment plant and system control operators. Their responsibilities are to carry out the operating plan.

The OPS develops the daily operating plan using data from

- a Maintenance Scheduler,
- a Water Demand Forecaster,
- a SCADA system (including equipment and facilities information); and 4) an Energy Cost Scheduler; as well as water quality operating constraints and the Utility's Performance Criteria (information about how the company's management wants work performed).

Operations, in the middle of the figure, is the traditional function carried out by the operators. They receive the daily plan from the OPS and work with everyone in the utility (e.g., treatment plant operators, OPS) to ensure the plan is followed. A major part of the operators' duties is to compare the actual operation of the water system to the daily plan. Operators monitor information about the current operating state of the water system using the SCADA system. To help ensure the water system follows the plan, operators use flow control valves and other supervisory control capability through the SCADA system, to guide the actual operation of the water system according to the daily operating plan. Operators also deal with emergencies and take action to respond to emergencies that supersede the daily plan.

The Treatment Plants and the Pumping Plants are at the far right of the figure. Treatment Plant operators receive the daily operating plan directly from the OPS. Pumping plants receive the plan via a pump control program. Operators use Supervisory Control, through the SCADA system, to control pumping plants and equipment in the distribution system to guide operation of the water system in accordance with the operating plan. They receive information from the treatment plants, pumping plants, and the distribution system via the SCADA system. SCADA data and other operating information is also provided to the OPS.

DESCRIPTION OF COLORADO SPRINGS WATER SYSTEM

Colorado Springs Utilities is a multi-service utility providing potable and non-potable water, wastewater, gas, and electric service to a population of approximately 450,000 people in the Greater Colorado Springs, Colorado area. The gas system is primarily a distribution provider of purchased natural gas. The electric system includes generation (one gas, two coal, and three hydroelectric plants), transmission, and distribution. The Colorado Springs Water System is primarily a surface water source system. Less than one percent (1%) of the water supply comes from groundwater.

Surface water supply delivery is accomplished through a complex system of diversions, reservoirs, pipelines, and pumping stations. The system is supplied by eleven sources of water whose annual average potable use is approximately 85,783 acre-feet.

The water supply delivery currently terminates at four supply points for the distribution system including five CSU-owned treatment plants (Pine Valley, McCullough, Mesa, Ute Pass, Woodmen) and the Fountain Valley Authority Treatment Plant (partially owned by CSU). The existing capacity for these six sources, based upon treatment plant capacities, is 232 million gallons per day (MGD). These water sources vary in quality (especially fluoride concentrations, which are relatively high in the local watersheds) and cost of delivery. Additionally, water storage rights, diversion priorities, and exchange agreements mandate careful management and use of both local and transmountain water (water diverted over or through the continental divide from the Pacific drainage into the Atlantic drainage).

Colorado Springs' water distribution is generally a gravity system extending about 11 miles from west to east and 16.5 miles from north to south. Areas not served by gravity are generally served by pumping. Within the water service area, ground elevations range from about 7,800 feet in Cedar Heights to 5,750 feet in the southern part of the city.

Because of the wide range in ground elevations, there are five different major service levels:

1. Briargate
2. Templeton
3. Northfield
4. Highline
5. Lowline

Each lower service level is fed through pressure reducing valves (PRVs) and flow control valves from the higher levels and from its own distribution storage tanks. The distribution system contains over 1,700 miles of mains.

The static hydraulic gradient for each service level is established by the overflow elevation of storage facilities or by the hydraulic gradient elevation of pressure reducing valves (PRVs) downstream of the service level. Service level boundaries are established to maintain acceptable minimum distribution system pressures, which CSU currently has established at 60 pounds per square inch (psi). Water flowing between service levels is monitored and controlled by over 200 PRVs, more than 70 of which are connected to the Advanced Integrated Supervisory Control and Data Acquisition (AI SCADA) system.

All treated water is stored in covered reservoirs and tanks. Currently 36 treated water storage tanks and reservoirs occupy 28 sites with a combined capacity of 101.4 million gallons of water.

From an energy management perspective, the primary controllable components are 18 raw water pumps in four raw water pump stations and 75 potable water pumps in 25 potable water pump stations. Additionally, CSU has 33 Megawatts (MW) of hydroelectric generation capacity in its raw water delivery system. Twenty-eight MW of that is at the Tesla hydroelectric generation plant, which is directly upstream of the two largest water treatment plants (Pine Valley and McCullough) in the water system. Production of hydroelectric power is directly influenced by water system consumption and operation.

From a water quality perspective the primary controllable components are surface water sources into the water treatment plants, the water treatment plants themselves, and distribution system rechlorination facilities.

A Water Operations Control Center was established in the spring of 1980. The heart of the Control Center is the computerized water control system, the AI SCADA, controlling the water distribution and wastewater collection systems. By making adjustments through the computer, the operator can change valve positions, increase or decrease pressures, control pump operations, and maintain a steady flow of water and adequate storage throughout the system.

The AI SCADA system enables a high level of operational control while collecting real-time data critical to the utility business needs, including pressures, flows, raw and treated water storage, stream stage, pipeline releases, etc. Among the advanced applications enabled by AI SCADA are water accounting, system-wide control, business process workflow, and the EWQMS.

OVERVIEW OF THE CSU EWQMS

Figure 1.2 provides a functional overview specific to CSU's EWQMS. The generic EWQMS model in Figure 1.1 was modified to fit the business needs and leverage the unique multi-utility structure of CSU (e.g., electric, gas, water, wastewater, and non-potable utility systems). CSU's EWQMS consists of software and people whose function is to optimize the operation of the water system for the benefit of CSU. The following overview represents the EWQMS vision at CSU. Software and OPS processes will be implemented in sequential phases with less complex solutions brought on-line first. Software specifications defined in Chapter 4 and attached in the appendices were written for the EWQMS prototype phase.

Operations Planner and Scheduler (OPS)

The core of the CSU EWQMS is the OPS who develops the daily operating plan for the entire water system. The OPS develops the daily operating plan using data from:

- a Maintenance Scheduler;
- a Water Consumption Forecaster;
- a Water Source Analyzer
- a Water Quality Analyzer
- a SCADA system (including equipment and facilities information); and
- an Energy Cost Scheduler, as well as water quality operating parameters, and the Utility's Performance Criteria (information about how the company's management wants work performed).

The OPS also considers labor costs required to operate and maintain the system. In Figure 1.2, the OPS and Operations are shown within a dashed line because this function will be served through one organizational group. The existing system control operators will be involved in the development of the operating plans. The OPS is required to provide more technical support to the System Control Operator to enable optimum operations of the water system.

At CSU, the OPS is a single person who coordinates with multiple groups within the utility. The OPS's primary mission is to develop a daily operating plan. The daily operating plan contains the operating schedule for the water and wastewater systems. Storage targets ensure appropriate flows from transmission through treatment to distribution. The schedule shows which pumps must run, how valves are set, how much surface water is used, etc., for each time period in the day. Everyone who affects the operation of the water and wastewater system uses this plan.

The daily operating plan is developed to ensure the water system efficiently meets the customers' demand for water while complying with water quality/supply parameters. Developing such a plan is typically a complex task involving several conflicting requirements and requires computer-based analysis. To handle this complexity, the OPS uses the EWQMS software to prepare the plan. The software schedules production from raw water resources and treatment plants. It also schedules pump stations, transfer valves, and collection pumping. Deliveries to large customers are also scheduled based on time-of-use costing.

The OPS uses an optimization algorithm in the EWQMS to determine the production, transfer and treatment schedule based on forecasted consumption. A system simulation model is run to evaluate system operations and boundary conditions (e.g., reservoir limits – overflows and fire storage). If boundary conditions are not violated, an optimum solution is reached. If boundary conditions are violated, a new operations schedule is developed and the simulation model is rerun. This process is repeated until a daily plan satisfies all the constraints (e.g., water consumption, water quality, water supply, and limitations of the water distribution system).

Practically speaking, the OPS does not develop a plan for each day, one day at a time. Rather, several days, possibly a week or two, are used for the study period in preparing the operating plan. This happens because the mathematical techniques employed in the software programs work best when there are several days over which an optimal schedule is developed. As a result, daily plans are developed from these longer-term schedules prepared using optimization techniques. New schedules are developed if there are significant deviations of the water system from the daily plan. Typically, the OPS has several reasons to re-develop the plan every day (e.g., unscheduled events such as major equipment outages since the plan was developed, changing weather conditions).

Operations

Operations, in the middle of Figure 1.2, is the traditional function carried out by the system control operators. They know the daily plan because they receive it from the OPS. The Operators work with everyone in the utility to ensure the plan is followed. A major part of their duties is to compare the actual operation of the water system against the daily plan. They do this by monitoring information about the current operating state of the water system via the AI SCADA system. To help ensure that the water system follows the plan, they use zone transfer valves, pumps, and other supervisory control capability through the AI SCADA system to guide the actual operation of the water system according to the plan. In addition, they may also take action to respond to emergencies that supersede the daily plan.

The Operations function is the traditional SCADA function CSU uses to monitor and control the water and wastewater systems. Monitoring the systems is accomplished by the data acquisition portion of a SCADA system and control is accomplished using the supervisory control portion of SCADA. Adding a set of advanced applications software, such as EWQMS programs, and integrating them with SCADA resulted in the designation of CSU's system as an Advanced Integrated SCADA system.

System control operators use the AI SCADA system to carry out the utility's core business (i.e., delivering high quality water to Colorado Springs Utilities customers). The addition of an EWQMS to a SCADA system is intended to automate some tasks normally carried out by system control operators and others, and perform new tasks needed in the near future (e.g., scheduling power with CSU Electric). The OPS function provides the interface with CSU Electric, and participation in Load Forecasting, Generation Scheduling and Load Shedding processes. Coordination of forecasted water consumption/flow demands, Tesla reservoir maintenance activities, and changes in pipeline topology support the maximization of power production and more responsive operation of the Tesla hydroelectric facility. At 28 MW, Tesla is CSU's largest hydrogenerator. Additionally, the OPS function supports the accurate assessment and availability of load shedding through system information sharing, coordination, and planning.

Because the daily operating plan is the lowest cost hour-to-hour operating plan that conforms to all water quality constraints, which if followed would minimize the utility's operating cost, and any deviations from the plan would result in increased overall cost of operating the water system, operators work to ensure that the plan is followed. This is very important considering the volatility of electric prices. This includes coordination with CSU staff, changing the position of zone transfer valves, and using supervisory control, through the AI SCADA system, to guide the moment-to-moment operation of the water system and in accordance with the plan. When the actual operation of the water system drifts off course, operators work with the OPS to come up with a new plan. To do this, the OPS once again uses the system scheduler. Normally, these variations from the plan tend to be small. Larger changes called events require operators to use their experiential understanding of the system to respond appropriately.

Events are defined as something that occurs, or happens, at a point in time but takes virtually no time to happen. Major equipment failures, natural disasters, and sabotage are examples of events. Some events (e.g., widespread power failure, extreme weather days, major equipment failures) are known to occur. It is not a surprise that they occur – only when they occur. The OPS has the tools to develop response plans for each category of event. Then, when an event occurs, the operator's use both their knowledge of the water system and the appropriate plan prepared ahead of time by the OPS.

The daily operating plan may be suspended when an event happens. When an event occurs, operators use the appropriate plan prepared by the OPS and have guidelines to use during the event. Afterward, the OPS can assess actual performance against the plan and modify the plan as necessary.

Water and Wastewater System

At the far right of Figure 1.2 are the Raw Water Supply, Water Treatment Plant, Pump Stations, Distribution, Customer, Collection, Wastewater Treatment, and Non-potable Utility Components. Raw Water Supply, Water Treatment Plants, and Wastewater Treatment Plants receive the daily operating plan from operations. Pump Stations, the Distribution System, the Collection System, and in some cases the Customer (when the utility and the customer have special operating agreements, for example) receive commands from AI SCADA. When necessary, System Control Operators use Supervisory Control, through the AI SCADA system, to manually control pump stations and transfer valves in the distribution system to guide operation of the water system. They receive information from raw water supply, treatment plants, pump stations, distribution system, non-potable system, and customers via the AI SCADA system. AI SCADA data information is also provided to the OPS.

Water Quality Analyzer

The Water Quality Analyzer is a rules-based software program that receives data from a number of sources (e.g., manually taken field samples, lab samples, and SCADA data) and compares this data to water quality regulations and CSU's Performance Criteria. By constantly monitoring water quality across the Utility, the OPS and other users receive information about the entire water system (i.e., their analysis of water quality is from the perspective of the entire water system and not from one isolated point within the system). The Water Quality Analyzer will generate operating parameters and alarms to the OPS for use in developing the operating plan.

Historical data and tools including the System Simulator (See Appendix E) and/or a hydraulic model are used to test alternative solutions. "What if" scenarios are tested to find the best practicable solution. These solutions typically involve the Utility's Performance Criteria, operating plans, and possibly investments in equipment. When an accepted solution involves changing operation of the water system because of water quality parameters, these new procedures become part of the Utility's Performance Criteria used by the OPS in preparing the daily operating plan.

It is not within the domain of the OPS to resolve any apparent conflicts between water quality regulations and energy cost minimization. Water quality regulations and the "Utility's Performance Criteria" are inputs to EWQMS via the Water Quality Analyzer. Energy optimization is performed within the constraints thus defined. When the water quality regulations and/or the utility's performance criteria change, the effect is to redefine the constraints within which the OPS must operate to find the least cost schedule for water system operations.

Any seeming conflicts between water quality constraints and energy optimization will be decided outside of EWQMS. Management will set the priorities. For example, assume the Mesa Treatment Plant is having problems with high fluoride water. A solution may be to restrict one (or several) of the local water sources into Mesa. Such a solution would effectively increase the allocation of booster pumps and increase the cost of energy. If this solution is adopted, the OPS would adjust the Treatment Plant Production Schedule and develop a new pumping schedule with the System Scheduler.

So the link between water quality and energy cost minimization is decoupled by having water quality staff and OPS work with management. Their decisions change the Utility's Performance Criteria, which are used by OPS to develop the daily operating plan. As a result, all operating plans are developed to be the least costly way to operate the water system while recognizing water quality operating parameters.

Water Source Analyzer

Water rights significantly influence the operation of the Colorado Springs Water System. There are twelve developed raw water sources ranging from 700 acre-feet (AF)/year to 35,000 AF/year. The utility must consider water law, water rights, priorities, historical operating data, performance criteria, hydroelectric generation potential, and water quality when establishing the water supply schedule and raw water pumping rules. Managing water resources is highly complex and changes frequently based on consumption, weather, and utility operations and maintenance. The Water Source Analyzer is a rules-based software program that evaluates all of the previously defined input data and generates an optimum water source schedule and raw water pumping rules for the OPS.

Energy Cost Scheduler

CSU's Electric System Control Center will perform energy cost scheduling for the utility. The Electric System Control Center will provide the daily energy cost schedule to the OPS. Currently, the schedule is a two-tiered (Peak/Off peak) Time-Of-Use (TOU) rate structure. In the future a real-time pricing schedule, which more closely conforms to the value of energy, will be communicated from the Energy Cost Scheduler to the OPS. This will enable the utility to respond to energy supply and market conditions to the advantage of the utility. For example, should the utility require additional energy due to an incorrect energy demand forecast, faulty water consumption forecast, or an equipment failure, it may be advantageous to reduce energy used for pumping rather than to attempt to meet the additional demand by purchases on the (volatile) spot market.

The Energy Cost Scheduler will also work closely with the OPS to develop an optimized schedule for the hydroelectric generation plants. An interruption schedule for energy conservation, and potentially, a signal for immediate shutdown of pumping is also generated from the Energy Cost Scheduler.

Ancillary Functions

Ancillary means in support of or integral to. In this case, it means applications software programs supporting the functions of the EWQMS. These programs make it possible for the OPS to develop the daily operating plan.

A Consumption Forecaster provides the projected demand for water in the future. This forecast is used to schedule the treatment plants and ensure that enough water is produced to meet the customers' demand. It is also used to schedule pumping to ensure pressure and flows are within guidelines. A Consumption Forecast is developed for the next day or days with one-hour resolution for each service area in the distribution system.

The Maintenance Scheduler prepares a master schedule of all planned equipment maintenance. This does not include unplanned outages, which are handled as events by the operators. The Maintenance Scheduler receives all planned maintenance schedules from various groups in the Utility. It resolves any conflicts (e.g., scheduling so many pump repairs at the same time that the system cannot reliably meet demands without violating the utility's performance criteria). The master schedule is distributed to all interested parties once the scheduling conflicts have been resolved. A clearance request system using CSU's Intranet is used to schedule maintenance affecting operations of the water system. The Maintenance Scheduler uses the consumption forecast, as well as other relevant schedules (including previously approved clearances), to decide whether it is prudent to approve a clearance request.

OBJECTIVES OF THE CSU EWQMS

The following objectives of the CSU EWQMS were developed from the benefit analysis and were not prioritized or reconciled:

1. Minimize the cost of pumping
2. Maximize the value of hydroelectric generation
3. Optimize the quality of water delivered
4. Efficiently/effectively utilize water supplies
5. Minimize the cost of treatment
6. Respond efficiently to sudden changes in operating conditions
7. Maintain high system reliability
8. Minimize the cost of maintenance
9. Maximize employee safety and security through planning and scheduling work on the water system

10. Maximize efficient and effective utilization of resources through formal operations planning and scheduling
11. Optimize utilization of system storage
12. Maximize system availability through collaborative planning
13. Facilitate alignment and coordination of the organization's operational processes
14. Collaborate with Electric System Operations for mutual benefits (e.g., electric load management, energy cost reductions, load shedding)

ORGANIZATION OF REPORT

This report contains a description of CSU's approach to prototyping an EWQMS. The following is an overview of each chapter:

Chapter 2: Benefits of an EWQMS at CSU, defines the methodology for analyzing the benefits of an EWQMS. The results of this analysis are quantified. Both tangible and intangible benefits are presented. Additionally, the approach used to convince CSU management of the value of the project is defined.

Chapter 3: Operations Planning and Scheduling, is the concept of the water system operations planning and scheduling function presented in this section. It includes a description of the methodology used to develop the requirements and processes for operations planning at CSU.

Chapter 4: Specifications for EWQMS Software, is an overview of the EWQMS software system. Individual specifications for software applications are included in Appendices.

Chapter 5: Procurement, Development and Installation of EWQMS Software, defines the process used to procure and develop the EWQMS software. Issues associated with software development are also defined.

Chapter 6: Findings and Recommendations, documents results achieved by the project and lessons learned. Recommendations for any water utility contemplating an EWQMS are presented.

Appendix A contains water quality sample data used to evaluate the requirements for the Water Quality Analyzer.

Appendix B contains the Job Description for Operations Planner and Scheduler.

Appendix C contains the Functional Specification for a Consumption Forecaster.

Appendix D contains the Functional Specification for an Economic Pump/Valve Controller.

Appendix E contains the Functional Specification for a Water System Simulator.

Appendix F contains the Functional Specification for an Optimum Pump Scheduler.

Appendix G contains the Functional Specification for a Water Quality Analyzer.

Appendix H contains the Functional Specification for a Water Source Analyzer.

Appendix I contains the Functional Specification for a Daily Operating Plan.

Appendix J contains the Functional Specification for a Daily Operating Plan NetConference application.

Appendix K contains the Functional Specification for a Construction/Maintenance Clearance process.

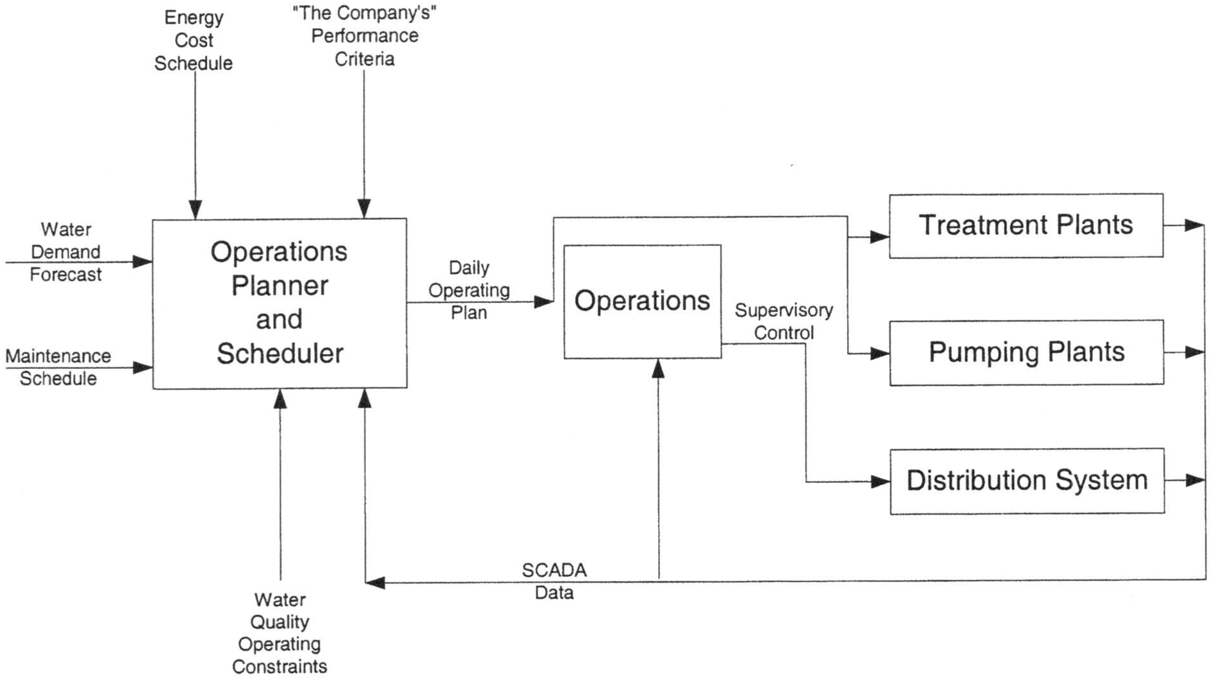

Figure 1.1 Functional overview of a generic EWQMS

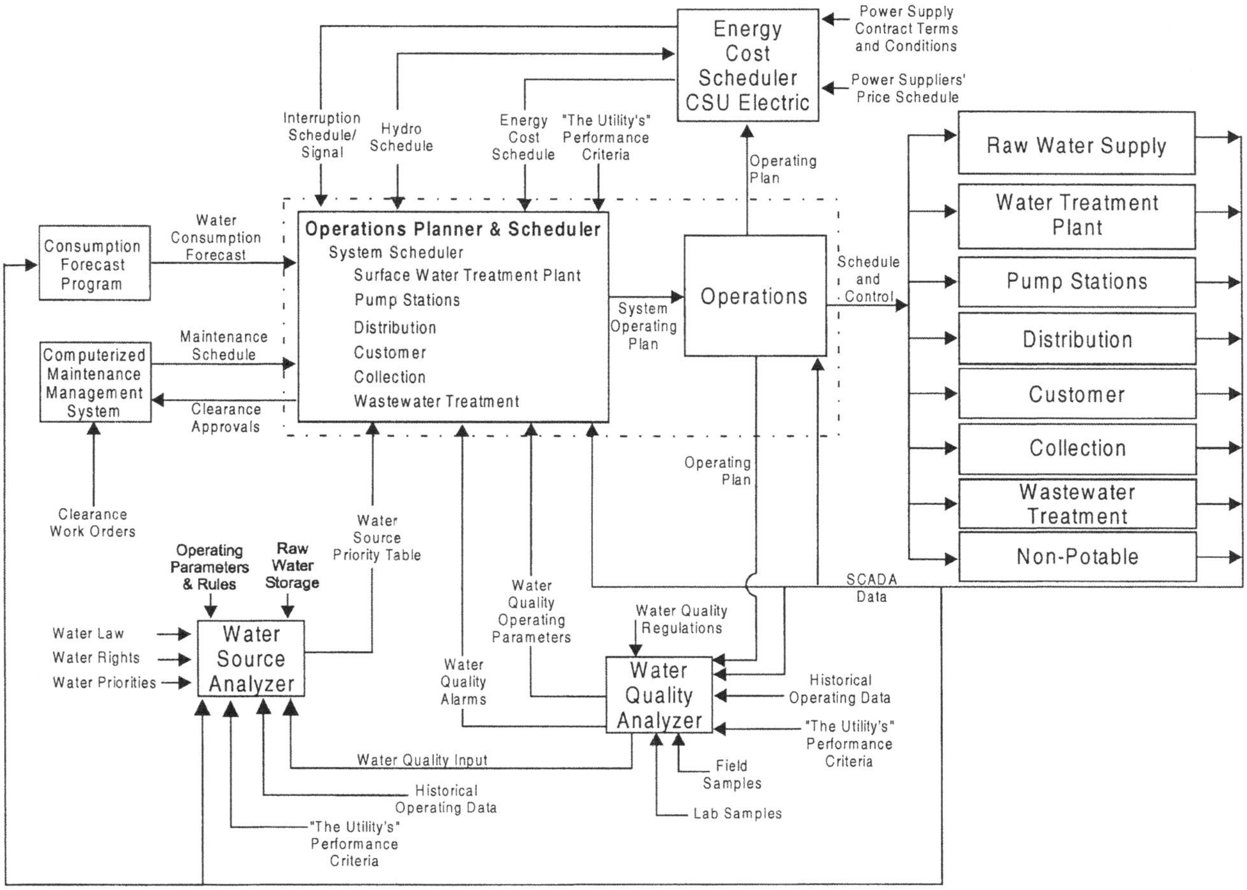

Figure 1.2 CSU EWQMS Model

CHAPTER 2
BENEFITS OF AN EWQMS AT CSU

In this chapter of the report, the benefits analysis of an EWQMS at CSU is presented.

Analysis of the benefits associated with an EWQMS initially focused on the supply and demand side of energy and water quality. In the process of modeling the functions of the OPS, other tangible and intangible benefits associated with the implementation of an EWQMS were discovered. Significant opportunities for system performance improvement were found by optimizing water source scheduling and water treatment plant scheduling.

BENEFIT ANALYSIS – ENERGY MANAGEMENT

The Energy Management benefit analysis focused on evaluation of energy cost reduction associated with shifting pumping load to off-peak hours considering CSU's two-tiered Time-of-Use (TOU) electric rates and increasing the production of the Tesla Hydroelectric Generation Plant.

To approximate the value of the pump optimization scenario, six different consumption days were selected for analysis in calendar year 1999:

Typical Winter Day	January 8, 1999
Minimum Winter Day	January 24, 1999
Spring Day	April 1, 1999
Typical Summer Day	July 30, 1999
Maximum Summer Day	July 13, 1999
Fall Day	November 17, 1999

The data from the analysis was then extrapolated for one year to estimate the potential benefits.

Benefit of Shifting Pumping Load to Off Peak Hours

CSU has a total connected electrical pumping load of approximately 23 megawatts. Connected load is a summation of all pumps in the CSU water system. Most of this pumping load (66%) is in the raw water system. Electric rates, hydraulics, and supply restrictions limit distribution system pumping as the only "controllable" load from an energy management perspective.

To calculate the distribution system aggregate electrical load profile, pump run times (one hour resolution) for all distribution pumps was retrieved from historical SCADA data. (CSU did not have hourly energy data metered at the pump stations.) Based on run time and pump horsepower ratings, estimated energy consumption for each pump was calculated on an hourly basis. Aggregate water system load profiles for the six days are shown in Figures 2.1 through 2.6.

To move pumping to off-peak times, the total run time for each pump was held constant for the day but shifted to off-peak hours where possible. Figures 2.1, 2.2, and 2.6 illustrate a reduction in pumping from 16:00-22:00 hours for the winter days and Figures 2.3 through 2.5

show a reduction in pumping from 11:00-18:00 hours for the summer days. This corresponds to the off-peak hours for the seasonal TOU rate schedule 500KWE8T. The on-peak cost of energy is $.0471/KWH and the off-peak cost of energy is $.0200/KWH. Demand charges were also included in the analysis. Results of the pumping analysis are shown in Table 2.1. The calculated benefit of moving pumping to off-peak hours was calculated from the difference between the actual aggregated energy and demand changes and theoretical case of shifting energy consumption to off-peak hours. The annual savings is an estimated $30,318 reflecting a 15% savings in energy and demand costs over the actual performance of the water system. If there were no hydraulic or storage restrictions and all pumping was moved to off-peak hours, the annual savings is an estimated $42,759 or 22%. Potential savings range between 15% and 22%.

The relatively small magnitude of savings is the result of two factors:

1. Energy is relatively inexpensive in Colorado Springs
2. CSU's largest pumping electrical load, raw water pumping, is not considered in the analysis because it is non-controllable from an energy perspective

Table 2.1
Comparison of TOU Scenarios

	1999 Performance	**All Pumping OFF Peak**	**Maximum Practicable Pumping OFF Peak**
Actual Water Consumption, 1999	25,672 MG	25,672 MG	25,672 MG
Total Calculated Pump Energy Use	7,116 MWH	7,116 MWH	7,116 MWH
On Peak Pumping Energy	1,466 MWH	0 MWH	411 MWH
Off Peak Pumping Energy	5,650 MWH	7,116 MWH	6,705 MWH
Calculated On Peak Energy Cost	$69,048	$0	$19,361
Calculated Off Peak Energy Cost	$112,992	$142,312	$134,091
Calculated Total Energy Cost	$182,040	$142,312	$153,451
Calculated On Peak Demand Charges	$4,636	$0	$1,301
Calculated Off Peak Demand Charges	$5,577	$7,182	$7,182
Calculated Total Demand Charges	$10,213	$7,182	$8,483
Total Electrical Costs	$192,253	$149,494	$161,935
Savings	---	$42,759 *22%*	$30,318 *15%*

Benefit of Increasing Energy Production of the Tesla Hydroelectric Generation Plant

The Tesla Hydroelectric Generation Plant has a rated maximum production output of 28 MW. Figure 2.7 is a simplified schematic of the water system in the vicinity of the Tesla Hydrogenerator. The energy produced by this plant is the least expensive source of energy for the CSU electrical system. Water from Rampart Reservoir flows through the Tesla Generator on its way to the McCullough and Pine Valley Water Treatment Plants. The Pine Valley Pipeline from Rampart Reservoir and the Northfield Pipeline from Northfield Reservoir also supply the two plants.

Water flow through the Tesla facility is limited by the following factors:

- Water system consumption
- Controllable storage of the Tesla Reservoir
- Water that bypasses Tesla Generator through the Pine Valley and Northfield Pipelines
- Requirement to use as much "local" water as possible in lieu of Transmountain water that flows through the hydrogenerator (water rights constraint)

A reasonable approximation of the opportunity to increase output from Tesla was found by calculating the annual available production from the McCullough and Pine Valley Water Treatment Plants and routing all this water through the hydro plant less water that flows through Northfield and Pine Valley Pipelines. Water that flows through the Northfield and Pine Valley Pipelines feed the two water treatment plants and bypasses the Tesla Generator. Northfield Pipeline flow was held constant in all cases. Based on a year of actual production data, that CSU Electric calculated, using their Production Cost Model, the potential benefit of increasing Tesla output is $180,000 per year (Case 1 in Table 2.2). Reducing Flow through the Pine Valley Pipeline could add an additional $50,000 in generated energy from Tesla (Case 2 in Table 2.2).

Table 2.2
Value of increased production of energy at Tesla Hydroelectric Plant

		Base Case	Case 1	Case 2
Generation at Tesla Hydro	MWH	50,666	67,980	72,695
Water Thru Tesla Hydro	MG	11,088	15,163	16,216
Converting Factor	MWH/MG	4.57	4.48	4.48
Water Production, McCullough Water Treatment Plant	MG	7,237	7,237	7,237
Water Production, Pine Valley Water Treatment Plant	MG	10,677	10,677	10,677
Northfield Pipeline Flow	MG	1,277	1,277	1,277
Pine Valley Pipeline Flow	MG	3,631	1,474	422
EDS Flow and Unaccounted for Flow	MG	1,918		
Increased Hydro Production – Value over Base Case	Dollar		180,300	230,150

The initial analysis indicated there is a significant opportunity to increase production of the Tesla hydroelectric generation unit through improved water and electric system operations.

BENEFIT ANALYSIS – WATER QUALITY

The water quality benefits associated with an EWQMS focused on the most significant water quality issues that face Colorado Springs Utilities:

1. Reduction of water age in the distribution system
2. Reduction of low chlorine residuals in parts of the system
3. Reduction of fluoride concentrations leaving the Mesa Water Treatment Plant in order to assure that the fluoride notification limit of 2 mg/L is not exceeded
4. Definition of water quality data needs

Various analyses and preliminary modeling were performed in order to provide a qualitative assessment of potential water quality benefits associated with improved operations afforded by the EWQMS.

The most significant water quality issues in the CSU system are (1) relatively high water age and resulting low chlorine residuals in portions of the distribution system, and (2) high fluoride concentrations in the local watershed. These issues were being assessed in terms of how an EWQMS could result in improved conditions or reduce costs to the utility.

Potential benefits associated with reducing the incidence of low chlorine residual or old water are being assessed by examining the historical occurrence of low chlorine residual in the distribution system and investigating the aging situation in distribution system reservoirs.

Though improved operation of the system for water quality purposes may result in some direct dollar savings (e.g., reduced costs of chlorine) the primary benefits would accrue from a higher water quality product and reduced probabilities of not meeting water quality standards. Further refinement of this benefit analysis is anticipated with the completion of CSU's capability to perform extended period simulation and water quality simulation of their distribution system. Water Quality Laboratory personnel, who participated on the EWQMS I-Team, anticipate that this effort will allow them to predict potential instances of low chlorine residual or high fluoride concentration with enough accuracy and lead time to avoid excursions from the utility's performance criteria and/or water quality constraints.

Potential benefits associated with modified operations to assure meeting fluoride standards are investigated by examining historical fluoride concentrations at the Mesa Water Treatment Plant and review of recent tests conducted by CSU on reducing fluoride levels during the treatment process.

Water Aging and Loss of Chlorine Residuals in Storage Facilities

In most distribution systems, the majority of water aging occurs in storage facilities in the distribution system. The primary result of this aging is loss of chlorine residual. Loss of this residual can result in bacterial growth in the storage facility or in piping after the water has left the storage facility. Modifications in operation can reduce the time that water spends in storage and could potentially result in direct cost savings if the chlorine dosage was reduced at the treatment plant or in rechlorination facilities. Additional direct cost savings could result from a reduced need for flushing of distribution mains.

Seven representative reservoirs in the CSU system were selected, from the list in Table 2.3, for detailed assessment of water aging and loss of chlorine residual. These facilities were selected to represent a cross-section of facility types, sizes, and locations and because of the availability of data to support the analysis. Potential benefits associated with low chlorine residual or old water in storage facilities were assessed by the use of a mathematical model of the storage facilities.

Table 2.3
Summary of CSU tank/reservoir characteristics

NAME	VOL.	Overflow	Depth	Shape	Dia	Inlet	Outlet
	(MG)	Elev. (ft)	(ft)		(ft)	Dia-in.	Dia-in.
Austin Bluffs #1	10	6688	30	RECT			
Austin Bluffs #2	5	6688	32	CYL	163	24	24
Bear Creek	2	6584	32	CYL	104	12	16
Briargate	3	7135	32				
Broadmoor #1	0.22	6800	15	CYL	25	10	9,10
Broadoor #3	1.5	6467	16	CYL	92	6	8
Broadmoor #4	0.75	6629	24	CYL	75	12	12,12
Broadmoor #5	1	7019	19	CYL	92	6	8
Cedar Heights #1	0.25	6651	16	CYL	52		
Cedar Heights #2	0.25	7238	16	CYL	52	8	12
Cedar Heights #3	0.25	7938	16	CYL	52		
Cheyenne Canyon	3	6413	32	CYL	126	30	24
Ellicott #1	5	6413	32	CYL	166	16	12
Ellicott #2	5	6413	32	CYL	166	16	12
FVA Terminal	2.4	5982	40				
Highline	20	6413	30	RECT			
Highway 83	3	7135	25	CYL	143	24	24
Little Mesa	5	6225	32	CYL	165	20	24
Lower Neal Ranch	1	6629	30	CYL	75	16	16
McCullough	5	7168	24	RECT			
Northgate	2	6900	25	CYL	120	36	24
Pine Valley (2 tanks)	10	7078	24	RECT	160x180	42	
Research	5	6935	32	CYL	165	24	
Skyway #1	0.22	6452	40	CYL	30		
Skyway #2	0.2	6445	30	CYL	34		
Skyway #3	0.21	6447	24	CYL	39		
Sunrise	0.25	6858	13	RECT	65x49	12	12
Templeton	2	6900	32	CYL	104	16	16
Upper Neal Ranch	0.25	7015	30	CYL	36	12	12
Upper Woodmen Valley	0.5	7430	24	CYL	60	12	12
Wilson	5	7134	32	CYL	84	24	24
Woodmen Valley	2	7230	16	CYL	130	12	16

CompTank, a simplified reservoir model developed as part of the AwwaRF project *Water Quality Modeling of Distribution System Storage Facilities* (Grayman et al. 2000), was used in this assessment. This model assumes one of several different ideal mixing patterns within the storage facility and calculates the resulting water age and chlorine residual discharging from the facility. For this analysis, complete and instantaneous mixing was assumed (a common assumption in distribution system modeling). It was also assumed that the chlorine residual entering each tank was 1 mg/L. This assumption does not hold for many tanks in the system where their influent has traveled long distances or has previously been resident in other upstream tanks. Such effects will be studied using an extended period simulation (EPS) system model of the distribution system.

One or two years of hourly operational data for the facility were accessed for each of the selected tanks or reservoirs. This data included hourly measurements of water volume in the tank and/or hourly measurements of flow entering the tank. For a tank that operates in fill and draw mode, only one set of data (volumes or flow) was needed for the analysis. For tanks that have simultaneous inflow and outflow, both volume and flow data were required.

To support the chlorine modeling, CSU personnel conducted a bottle decay test on the water leaving the Mesa Treatment Plant. This test was used to determine the rate at which the chlorine will decay. Several water samples were taken of the effluent and stored in separate dark bottles at a constant temperature. Bottles were subsequently opened at different intervals and tested for remaining chlorine residual. The results were plotted and a rate coefficient fit to the data. Results of the analysis are shown in Figure 2.8. Generally the rate curve will plot as an approximate straight line on semi-log paper corresponding to a first order decay. However, as was the case for the Mesa water, sometimes the data does not fit a first order decay. Since first order decay functions are required by many models, as illustrated in Figure 2.8, a range of first order decay constants could be selected to fit the data. For the present analysis, a value of 0.30 per day was selected as the representative decay coefficient. This corresponds to a half-life (the time until chlorine residual decays to 50% of the original concentration) of 2.3 days.

Subsequent to the completion of the analysis described in this report, a second bottle test was completed. This test resulted in closer agreement to a first order decay function with a decay rate of 0.15 parts per million (ppm) per day. This corresponds to a half-life of 4.6 days. The result of the use of this lower decay rate would be less chlorine loss than is indicated in the reservoir analysis described below. However, it would not change the relative degree of loss among the reservoirs being compared in the analysis.

The results of the water age and chlorine modeling study for the seven reservoirs are presented in Appendix A and analyzed below.

Little Mesa Reservoir is a five million-gallon facility. It is located on a direct route from Highline Reservoir (Mesa Water Treatment Plant) and as a consequence, the age of the water entering Little Mesa is very low and the chlorine residual is not greatly diminished from the residual leaving the plant. One year of data (1999) was analyzed using the CompTank model. As an example, the resulting simulated water age and chlorine residual in the effluent from the reservoir is presented in Figures 2.9 and 2.10, respectively. As illustrated, in the summer, the residence time in Little Mesa is less than one day with a resulting loss of chlorine residual of around 15%. However, in the winter, the residence time climbs to three to five days and approximately half of the chlorine residual is lost during the time the water is stored in the reservoir.

Cedar Heights Tank #1 is a 0.25 million-gallon cylindrical reservoir. Water from this tank is pumped into Cedar Heights Tank #2 and subsequently to Cedar Heights Tank #3. As a result, the water in this portion of the distribution system can get very old. A rechlorination station adds chlorine to the water as it is pumped into Cedar Heights Tank #2. As illustrated in the graphs in Appendix A, during the summer the residence time in Cedar Heights Tank #1 is between 1 and 1.5 days. This increases to approximately three days in the winter. Chlorine loss is estimated at between 20% and 30% in the summer and between 40 % and - 60% in the winter, based on the model. The cascading effect as water goes from Cedar Heights Tank #1 to #2 and then #3 can result in very old water and deteriorating water quality. The rechlorination facility is designed to, in part, mitigate the impacts of the water-aging problem in this area.

Cheyenne Canyon Tank is a three million gallon cylindrical tank. The tank is located near a far end of the system so that the travel time to the tank is relatively high. This is one of the few storage facilities in the CSU system that is operated in a fill and draw mode (as opposed to simultaneous inflow and outflow). Additionally, there may be a "sloshing" action in the vicinity of the tank where the same water leaves and re-enters the tank. This problem is especially evident during the winter. Operators are quite conscious of the water age problem in the tank and are operating the tank so that it is drawn down significantly before it is refilled. Modeled residence time of water in the tank is generally between two and four days old and chlorine loss is between 40 and 60%.

Ellicott Tank is actually composed of two identical 5-million gallon cylindrical tanks. It is being modeled as a single composite 10-million gallon tank. It was modeled primarily from Pine Valley and McCullough Treatment Plants and there is a long distance and circuitous path from the plant to the tank. The tank is fed primarily through the Northfield and Templeton systems and, as a result, since water is relatively old reaching the tank, there is a rechlorination facility on the inflow to the tank. In the winter, residence time is between 1 and 2 days and chlorine loss is approximately 30%. In the summer, residence time drops to less than a day and chlorine loss to under 20%. All residence times, chlorine residuals, and water ages cited are those obtained from the modeling study, and should be interpreted as model data, not actual data.

McCullough Reservoir is a five million gallon reservoir located at the treatment plant. Residence time is typically three to 10 hours and chlorine loss is minimal (under 5%).

Highway 83 Reservoir is a three million gallon cylindrical reservoir at the extreme northeastern end of the distribution system. It is currently one of the most difficult operational situations in the system because of its distant location and the "sloshing" effect of water exiting and re-entering the tank. Residence time varies between 1 and 4 days and chlorine loss between 20 and 60%. This area of the system is developing quite rapidly and as this development occurs and demand increases, the residence time should decrease accordingly.

Highline Reservoir is a 20 million gallon reservoir located next to the Mesa Plant. Residence time in the reservoir is less than a day and chlorine loss in the range of 5% to 20%.

As illustrated by the analysis of the seven reservoirs, there are a wide variety of situations in the system relative to residence time and chlorine loss in the reservoirs. In general, except for the reservoirs located in close proximity to the treatment plants, residence times of several days are quite common with resulting chlorine losses of 10% to 60%. This certainly suggests the possibility of significant potential for benefits in the form of improved water quality and reduced chlorine usage through operational changes in the system.

Low Chlorine in the Distribution System

A related water quality problem is low chlorine residual in parts of the distribution system. This can result from long residence times in reservoirs, cascading of water from one reservoir to downstream reservoirs, long travel times in water mains, and loss of chlorine residual through interaction with pipe walls associated with corrosion and biofilms. The extent of this problem (and thus the identification of potential benefits through improved operation) can be determined through modeling or evaluation of sampling data. In this section, field data is examined.

In order to perform an initial assessment of the extent of low chlorine residual in the distribution system, all available chlorine sampling data for the years 1998 and 1999 (through

December 15, 1999) was examined. Data was available for approximately 75 stations sampling stations in the distribution system. There were 56 stations with at least 100 measurements available for the two-year period, and these stations were selected for analysis. The number of measurements varied from 103 to 322 at these stations.

For each station, the number of samples, the mean, minimum, maximum and standard deviation were determined and are presented in Table 2.4. Additionally, time series plots of chlorine residual at each station were prepared and are included in Appendix A. Stations with average chlorine residuals less than 0.3 mg/L were identified as the most likely potential problems. These stations are highlighted in bold in Table 2.4 and are briefly discussed below.

B06_102LEX is located at a school in a rapidly developing part of the system. It receives its water through Highway 83 Reservoir. Chlorine varies between 0.15 and 0.55 mg/L with lower residuals occurring during the winter. The longer residence times in the reservoir in the winter and relatively long travel times appear to be the primary causes of the depressed chlorine residuals.

B09A_27DUB is located in the back maintenance area of a pharmacy. As illustrated in the example plot in Figure 2.11, there is a very significant seasonal variation at this station with chlorine residuals dropping from 0.5 - 0.6 mg/L in the summer to 0.1 - 0.3 mg/L in the winter.

B14_86SCA is located at a school in a relatively distant part of the system. Relatively low (0.1 – 0.4 mg/L) chlorine residuals are observed throughout the year.

B34_21WCO is located at a convenience store in a relatively low volume consumption area. Additionally, there is cast iron pipe located in this area that may be reducing the chlorine residual.

B36_CHTK3 receives its water from Cedar Heights Tank #3. However, the chlorine residual is generally much lower than the chlorine residual leaving the tank (typically 0.2 mg/L at this station versus 0.5 mg/L). It is hypothesized that this station is not located on a main line and there may be significant loss in chlorine due to long travel times.

B49_0UPL is located in the Broadmoor area. The water generally comes from the Cheyenne Canyon Reservoir via some relatively complex pumping paths. Chlorine residual is especially low (0.1 – 0.35 mg/L) in the winter.

B55_34DRE is located in the bathroom of a bus barn in a business area. It is fed by Highline water that travels a long distance from the treatment plant. Average chlorine residual is around 0.3 mg/L with slightly lower concentrations in the winter.

B61_PPCC is located at the Pikes Peak Community College in a relatively isolated part of the system. Chlorine residual is almost always quite low (0.1 to 0.3 mg/L) at this station. When Fountain Valley is supplying water to CSU, this site sometimes receives water from that source.

B83_H83TK is located near the outlet from Highway 83 Reservoir. Even though water is rechlorinated coming into Northgate Tank that feeds Highway 83 Reservoir, the chlorine residual at this station frequently experiences chlorine residuals below 0.3 mg/L.

Analysis of the chlorine residual data routinely collected in the field provides a partial picture of the spatial and temporal variation of chlorine in the distribution system. It indicates that chlorine residuals are generally lower in the winter in most locations than in the summer. It also indicates that relatively low chlorine residuals (say less than 0.3 mg/L) are relatively common at many locations. It is very likely that even lower chlorine residuals could be found at other locations that have not been sampled. This information certainly suggests that some improvements potentially could be made in operations that would result in higher chlorine residuals. As a result, water quality benefits could be derived from improved operation. Application of a distribution model representing chlorine will provide a more complete picture of the variation of chlorine and the potential benefits that could be derived from modifications of the system operation. Such a model is currently being developed/applied and is described in more detail as follows.

Table 2.4
Summary of CSU distribution system chlorine data 1998-1999
(Stations with low average chlorine residuals (< 0.30 mg/L) are shown in bold)

STATION	# SAMPLES	MEAN	STD. DEV.	MIN	MAX
B01A_23RAM	303	0.49	0.09	0.20	1.00
B02_49CEN	120	0.49	0.08	0.20	0.70
B03_6GOG	119	0.44	0.07	0.15	0.60
B04A_4WDM	126	0.46	0.09	0.25	0.65
B05B_DRURY	127	0.38	0.12	0.15	0.75
B06_102LEX	**132**	**0.29**	**0.10**	**0.10**	**0.55**
B07_76NUN	296	0.42	0.10	0.10	0.70
B09A_27DUB	**128**	**0.27**	**0.15**	**0.05**	**0.60**
B10_32WWD	120	0.38	0.13	0.10	0.65
B11_38MBL	127	0.37	0.12	0.04	0.65
B12A_44BAR	121	0.39	0.11	0.15	0.60
B13A_46DAP	295	0.33	0.11	<0.05	0.60
B14_86SCA	**297**	**0.20**	**0.11**	**0.05**	**0.70**
B15_NGTK	144	0.35	0.13	<0.05	0.60
B17B_57CON	119	0.48	0.08	0.30	0.70
B18A_14NAC	111	0.39	0.10	0.20	0.60
B19_PAFB	322	0.49	0.08	0.30	0.70
B20_6BRI	121	0.48	0.09	0.10	0.70
B21_0SCR	105	0.48	0.08	0.25	0.70
B22_5NUN	105	0.47	0.11	0.05	0.70
B23_8NNV	103	0.60	0.07	0.35	0.75
B24_9ESP	120	0.54	0.10	0.20	0.75
B25A_17NWB	120	0.54	0.09	0.30	0.75
B26_UCCS	294	0.47	0.10	0.20	0.65
B27_36NCS	120	0.49	0.08	0.20	0.70
B28_28MSR	126	0.42	0.09	0.15	0.60
B29_2NNV	126	0.58	0.07	0.30	0.75
B33_14WCO	288	0.54	0.10	0.10	0.70
B34_21WCO	**135**	**0.25**	**0.15**	**0.05**	**0.75**
B35_3CLM	120	0.46	0.09	0.25	0.70
B36_CHTK3	**120**	**0.16**	**0.07**	<0.05	**0.55**
B37_CHTK2	120	0.50	0.11	0.15	0.75
B38_103UP	128	0.48	0.12	0.15	0.70
B39_IBEL	128	0.51	0.14	0.15	0.75
B40_15ARC	120	0.38	0.10	0.10	0.60
B42_25SCP	293	0.37	0.12	0.10	0.60
B43_12CYR	120	0.41	0.11	0.10	0.65
B44_0FIR	293	0.48	0.13	<0.05	1.00
B45_BMRL	120	0.43	0.13	0.10	0.65
B46_TCAS	134	0.68	0.12	0.20	1.20
B49_0UPL	**312**	**0.27**	**0.10**	**0.10**	**0.70**
B50_POUR	118	0.48	0.09	0.15	0.70
B51_7CYM	127	0.47	0.10	0.25	0.70
B52_52FAR	119	0.30	0.12	0.10	0.55
B53_DUAL	154	0.56	0.13	0.10	0.80
B55_34DRE	**119**	**0.29**	**0.09**	**0.10**	**0.50**
B57_4LKW	119	0.54	0.08	0.35	0.70
B58_21FTN	119	0.45	0.10	0.20	0.70
B59_28SCR	119	0.40	0.12	0.15	1.00
B60_SGSTD	120	0.43	0.12	0.10	0.70
B61_PPCC	**117**	**0.18**	**0.11**	<0.05	**0.65**
B62_FVTT	249	0.64	0.14	0.10	0.90
B63A_REDL	119	0.42	0.09	0.15	0.60
B83_H83TK	**107**	**0.26**	**0.14**	<0.05	**0.75**
BCR_TK_EFF	284	0.34	0.10	<0.05	0.75
BMR_GRG	166	0.46	0.12	0.10	0.65
BX_FWQ	487	0.34	0.21	<0.05	1.40
MAIN_FLSH	831	0.47	0.14	0.05	0.90

Distribution System Modeling

Water quality processes in a distribution system are quite complex resulting in water quality that varies both spatially and temporally throughout the system. In order to assess water quality in a distribution system under different operating conditions, a computer model that represents the time varying water quality is essential. CSU has recently completed a steady-state distribution model. CSU will extend this model to handle time varying situations (known as extended period simulation or EPS) and to simulate water quality. Representative seasonal days have been identified, and the model will be applied for these days to study the variation in both water age and chlorine residual throughout the system. Alternative operations can also be studied to directly identify water quality benefits associated with improved operations.

Fluoride Concentrations

According to Environmental Protection Agency (EPA) drinking water standards, fluoride concentrations in excess of 2 mg/L require notification of customers, and 4 mg/L is the maximum allowable concentration. In the CSU system, much of the so-called "local" water supply exceeds the 2 mg/L fluoride concentration. Therefore, in order to assure a fluoride concentration in the distribution system that is below the 2 mg/L notification level (an important objective of CSU), the high fluoride water must either be treated and/or blended with trans-mountain water that has a lower fluoride content. Water rights issues complicate the situation that requires CSU to utilize the available local water or else face the possibility of losing water rights for some of the trans-mountain water.

CSU has analyzed the fluoride concentrations reaching the Mesa Treatment Plant from various sources and has investigated the potential for reduction of fluoride concentrations in the treatment process. The analysis indicates that without blending, the fluoride concentration entering Mesa Treatment Plant would frequently exceed the 2.0 mg/L target. The treatment assessment shows the addition of more coagulants can lead to fluoride reductions in the range of 0.2 to 0.4 mg/L. Generally such a reduction would be sufficient to preclude the need for blending. Since there are energy costs associated with the blending process and chemical costs associated with the additional coagulation, tradeoffs exist between water quality, energy, and chemical costs in order to meet the fluoride standard. This indicates there are potential benefits associated with improved operation of the system to meet this standard.

Conclusions of Water Quality Benefit Analysis

The following conclusions were reached in the water quality benefits analysis:

- Residence times of several days in many reservoirs and cascading of flow from reservoir to reservoir can lead to significant water aging and loss of chlorine residual
- Long travel times, circuitous travel paths, and loss of chlorine in older cast iron pipes can further result in reduced chlorine residuals in parts of the system leading to the potential for bacterial growth
- Water age increases and chlorine residuals decrease during the winter
- Sampling data provides a partial picture of the water quality situation but enhanced modeling (extended period simulation and water quality simulation) is needed to fully

assess benefits and to study alternative operational plans to improve water quality and reduce energy costs in the distribution system

- It is possible to avoid exceeding the fluoride notification limit of 2 mg/L through relatively complex treatment processes and by carefully blending source water, but at the cost of increased chemical and energy costs.
- An EWQMS can lead to improved operations resulting in water quality benefits

Though improved operation of the system for water quality purposes may result in some direct dollar savings (e.g., reduced costs of chlorine due to lower average water age and chlorine decay in the system) the primary benefits would accrue from a higher water quality product and reduced probabilities of not meeting water quality standards. Further refinement of this benefit analysis is anticipated with the completion of CSU's capability to perform extended period simulation and water quality simulation of their distribution system.

BENEFIT ANALYSIS – WATER SOURCE SCHEDULING

Development of the requirements for the Water Source Analyzer led to the discovery of significant benefits. The Water Source Analyzer will optimize the scheduling of the various sources of supply to the CSU system considering water quality, treatment cost, pumping cost, direct flow value, storage value, hydroelectric generation potential, and water rights.

All of the previously mentioned variables will be used by the Water Source Analyzer and the OPS to develop an optimum schedule for the various water sources into the CSU water system. This holistic or systems approach to water source analysis and scheduling is expected to have significant tangible benefits to CSU. The tangible benefits include greater utilization of hydroelectric generation, improved water quality, reduced treatment costs, and maximum utilization of water rights.

Initially, the Water Source Analyzer was viewed as a component of the CSU EWQMS, which would provide the operating constraints of water rights to the OPS and System Scheduler. However, the research team discovered that optimization in the Water Source Analyzer, considering the previously mentioned variables, provided "up stream" opportunities for benefits far beyond the original constraints of water rights only.

BENEFIT ANALYSIS – WATER TREATMENT

As mentioned in Chapter One, CSU operates six water treatment plants, all of which include rapid sand (actually anthracite/sand) filtration. All references to treatment plants in this report include such filtration. Abrupt large system flow changes are very expensive for such water treatment plants. Water Treatment benefits of the EWQMS are in two areas – labor and process optimization. As with the other benefit descriptions, these benefits accrue from a "systems" approach to optimization. That is, optimization considering the entire water system from the water resources to the customer.

Implementation of the OPS concept and true system operations permits the water treatment plants to minimize the need for control room operators at each of the four main water treatment facilities. Remote monitoring and system planning and scheduling "frees up" control room operators for plant maintenance functions.

Abrupt large system flow changes are very expensive for water plants, and present potential water quality problems. An EWQMS would enable CSU to minimize abrupt flow changes and instead make small gradual changes to the treatment schedule. By effective scheduling CSU can reduce the operating costs of the treatment facilities in many ways:

1. Plant production is currently changed on the average of two to five times per day during the summer months. In the winter the flow changes are one to two times per day. Through effective planning this can be reduced to one or two gradual changes per day in the summer, and one or two changes per week in the winter.
2. Reducing filter washes by 20% at the four main treatment facilities is possible by minimizing flow changes. This reduces the need for backwash water, wash water pumps, recovery basin pumps, and re-treatment of backwash water. Filter washing is one of the most expensive functions in a water treatment facility. A routine filter wash averages $300 per wash. This is the accumulated cost of backwash water pumps, surface wash pumps or air wash system, recovery rake systems, recovery basin sediment pumps, recovery basin water recovery pumps, retreating water used for backwash, and operator labor.
3. Treatment chemical cost increases when plants have to respond to abrupt, large (10 to 30% of plant capacity), flow changes on a daily basis. The routine treatment costs of the four major treatment plants range from $18 to $23 per million gallons. When abrupt flow changes occur, plants must adjust chemicals to compensate for process upsets. This can amount to an increase of $2 to $3 per million gallons, depending on the current conditions.
4. Reducing backwash cycles increases the service life of plant equipment. The reduced pump cycling for additional chemical addition, re-supplying wash water tanks, pumping down recovery basins, and boosting pressure with air systems or water pumps increase the service life of the equipment. If plants are running at optimal flows with minimal flow changes, the plants can perform planned periodic maintenance and keep maintenance costs to a minimum. When equipment has to be run at a higher rate for longer periods of time, the maintenance curve increases. Equipment replacement is also necessary at an increased frequency. Additional man-hours are required to meet the increased maintenance demand.
5. Abrupt flow changes in water plants increase the risk of break-through or material getting through the plant filters into the distribution system. This is one of the key reasons flow changes should be made in water plants at a gradual rate. CSU plant operators have responded to this challenge by compensating with extra chemicals in the treatment process to insure that no breakthroughs occur. With the EWQMS, plants can make better decisions about chemical dosages and chemical system process balances due to gradual increases or decreases in plant flows. With more restrictive water quality regulations, this will become more critical. CSU estimates that improved water treatment plant scheduling has a potential savings opportunity of $600,000 per year.

BENEFIT ANALYSIS – SYSTEM RELIABILITY, SECURITY, AND SAFETY

Modeling the processes of operations planning and scheduling revealed other previously unrecognized and unquantified benefits; system reliability, security, and safety. Implementing formal processes for scheduling work through clearance procedures will result in fewer errors in the maintenance of the water system. This will improve the safety of field crews and minimize service interruptions in the water system. Maintenance work will be more carefully planned and coordinated through the OPS. To effectively schedule operation of the major facilities, processes, and equipment in a water system, it is necessary to keep track of the current state of each facility or piece of equipment and any changes to that state. This facilitates emergency response and minimizes the effects of breaches in security.

RETURN ON THE EWQMS INVESTMENT

The capital investment for EWQMS software and implementation services is estimated at $400,000 over a two-year period. Additionally, half of the OPS's time will be required to integrate EWQMS software applications into daily operations ($55,000 annual burdened salary cost). Considering these costs and the previously defined energy management tangible benefits, the Return-On-Investment (ROI) is conservatively estimated at four years. However, if the water treatment plant scheduling benefits are considered and/or the value of energy continues to increase, the ROI may be less than two years.

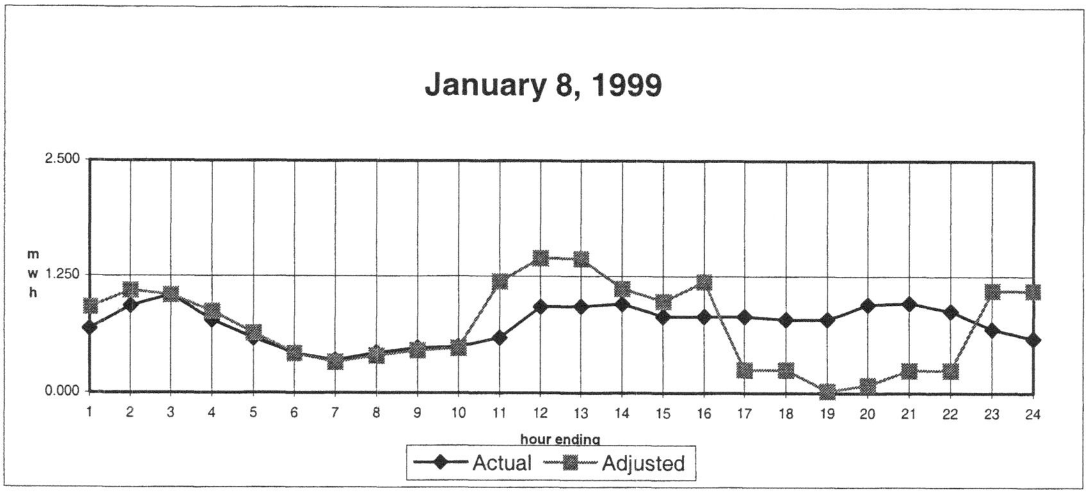

Figure 2.1 Load profile - minimum water consumption winter day

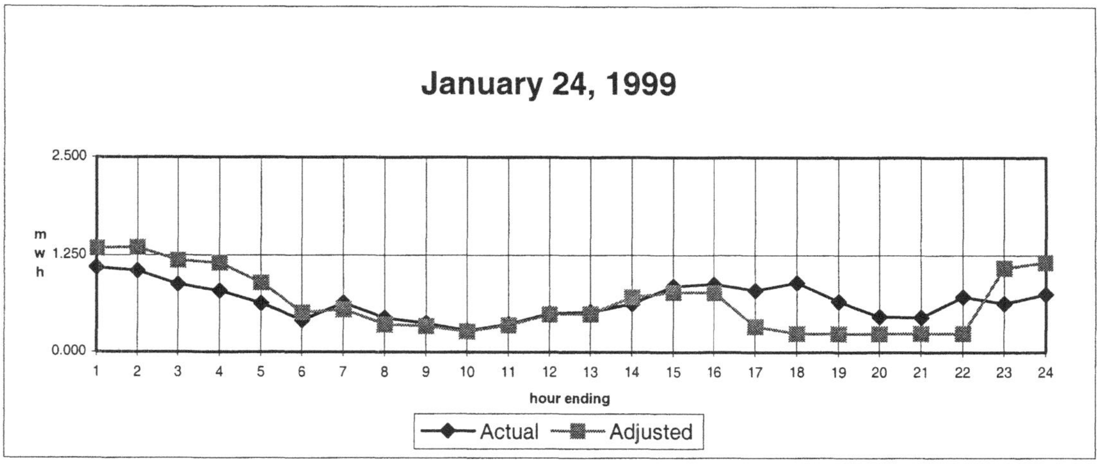

Figure 2.2 Load profile - typical water consumption winter day

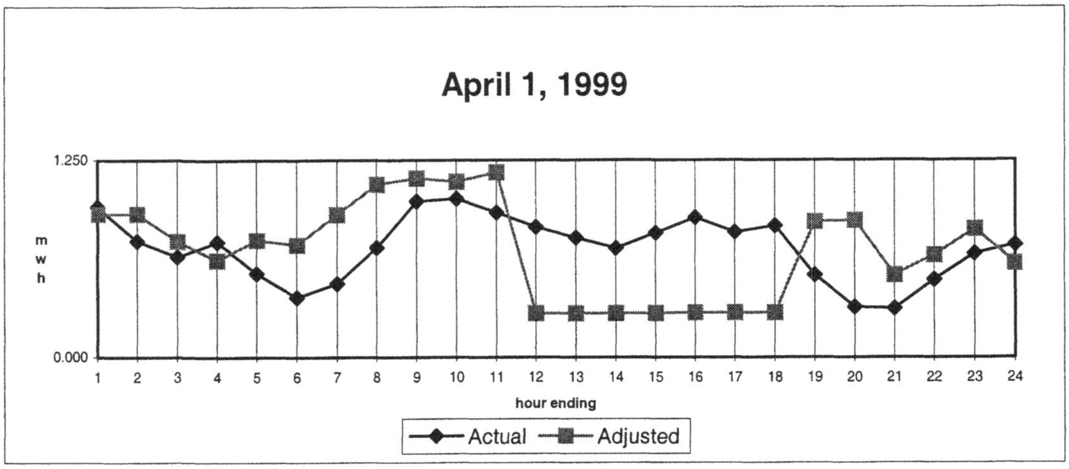

Figure 2.3 Load profile – typical water consumption spring day

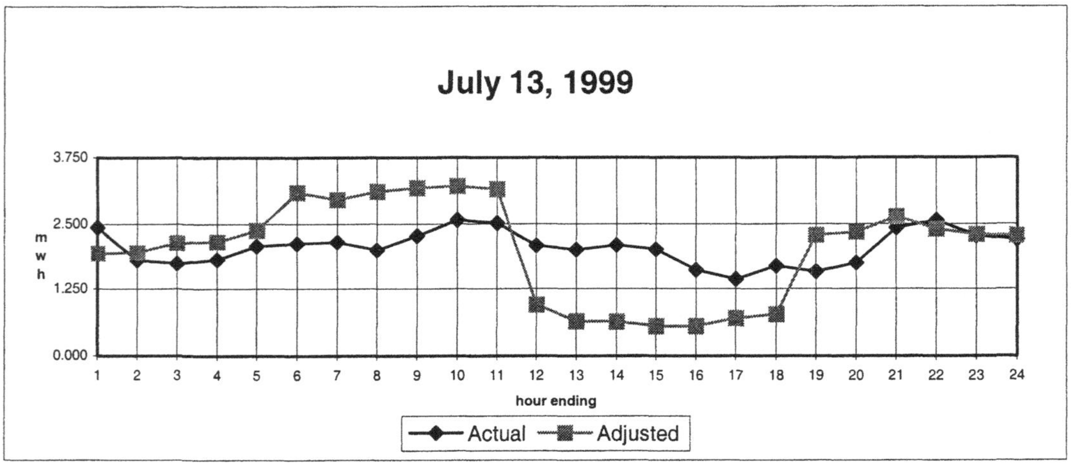

Figure 2.4 Load profile - maximum water consumption summer day

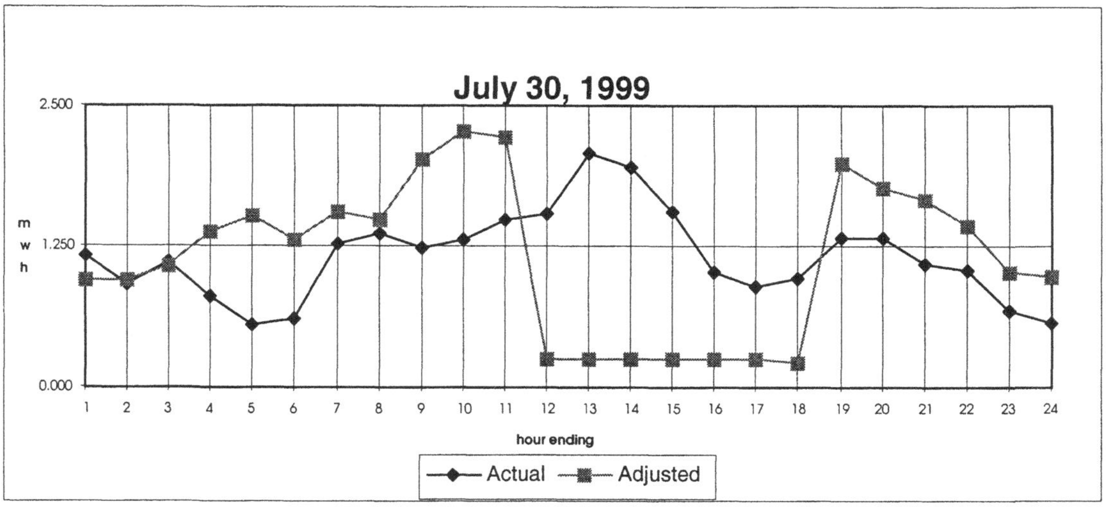

Figure 2.5 Load profile - typical water consumption summer day

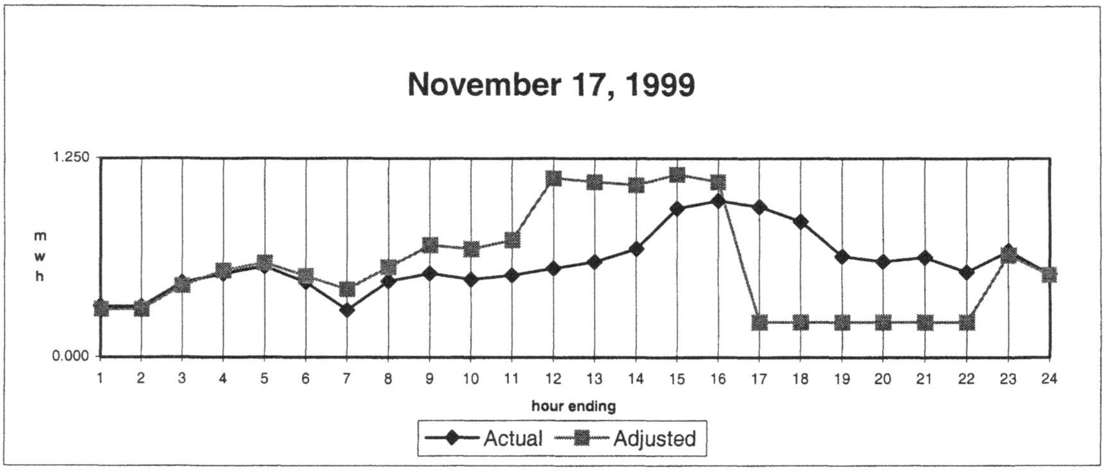

Figure 2.6 Load profile - typical consumption fall day

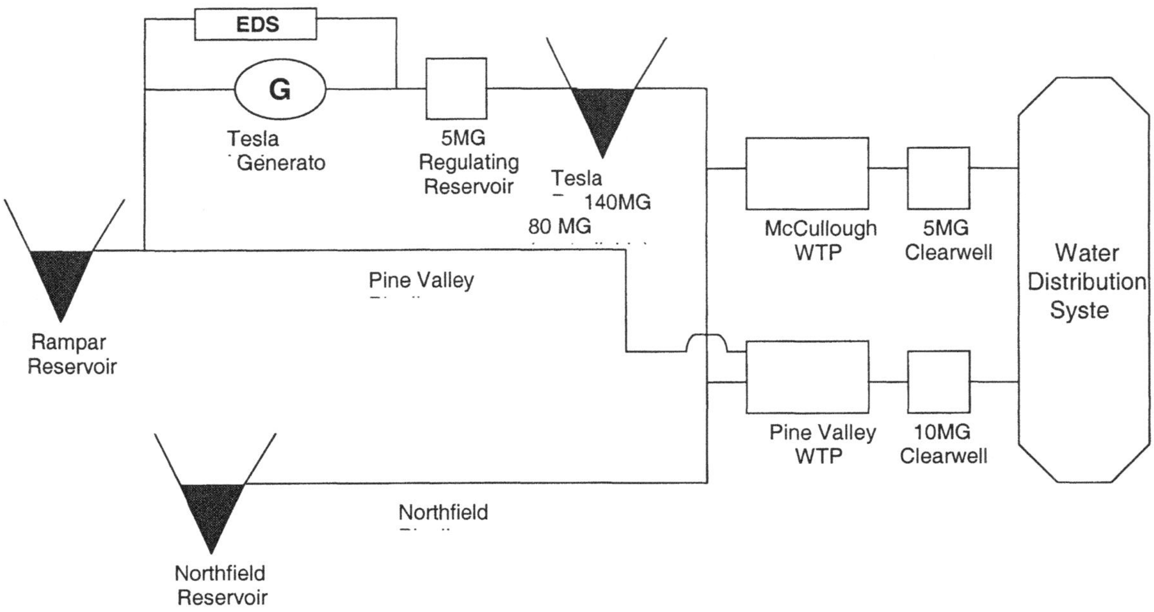

Figure 2.7 Simplified schematic – Tesla hydrogenerator and supply to McCullough and Pine Valley Water Treatment Plants

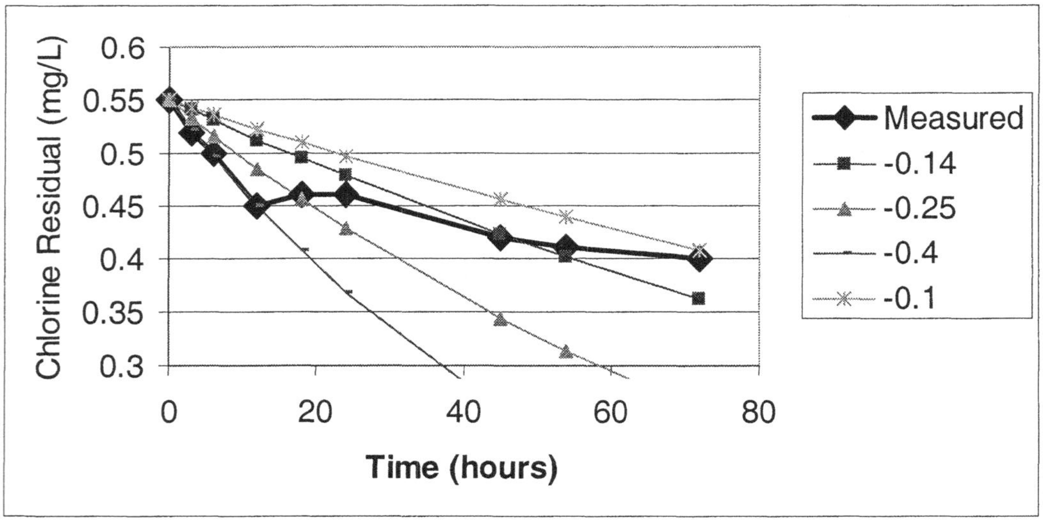

Figure 2.8 Chlorine bottle decay results

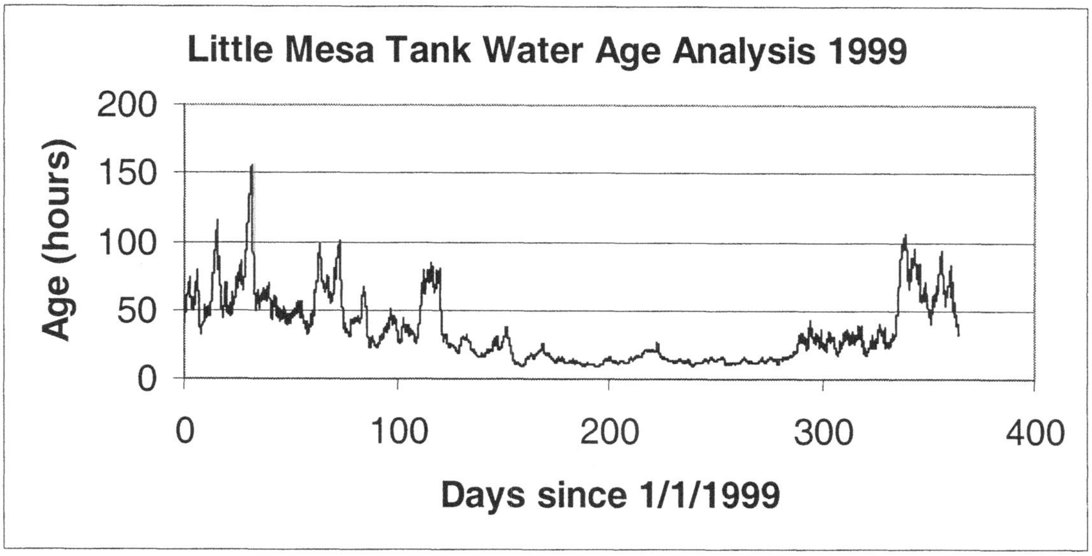

Figure 2.9 Modeled residence time in Little Mesa Tank

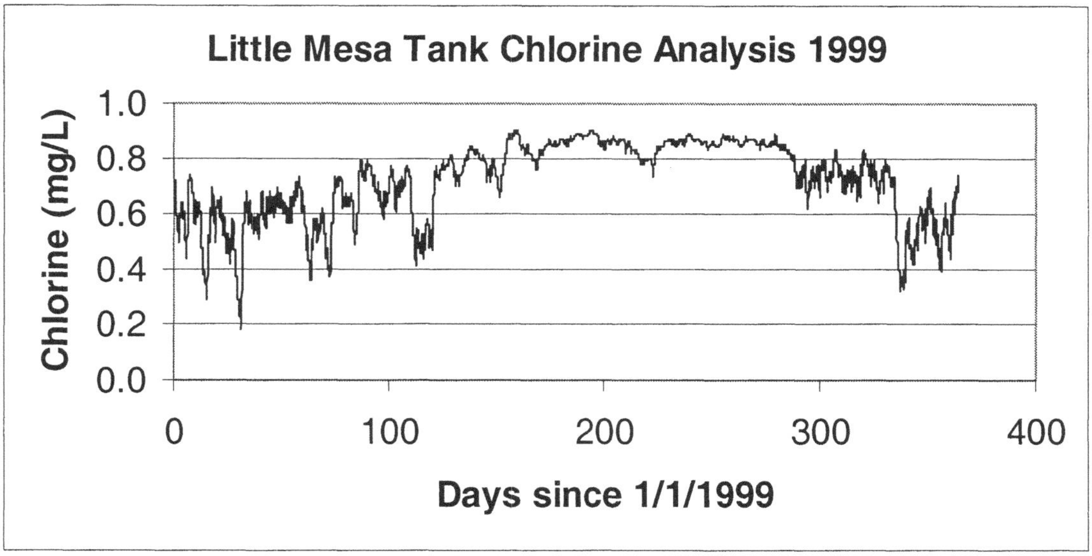

Figure 2.10 Modeled chlorine residual leaving Little Mesa Tank

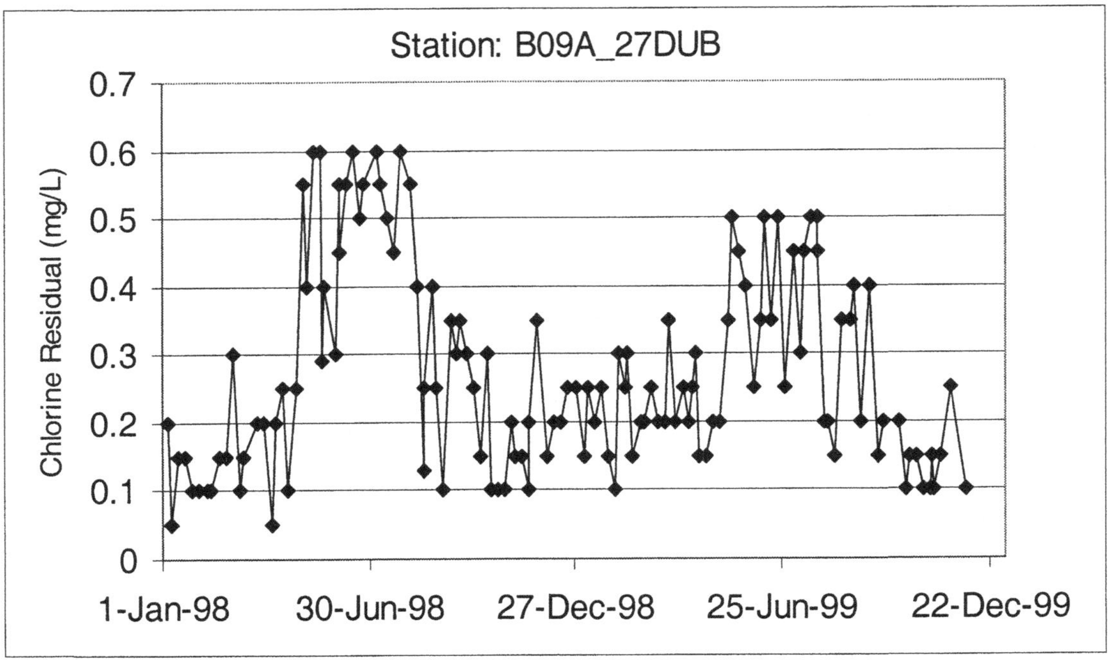

Figure 2.11 Example plot of sampled chlorine residuals

CHAPTER 3
OPERATIONS PLANNING AND SCHEDULING

In previous EWQMS projects sponsored by AwwaRF and EPRI (Jentgen et al. 1997), the notion of an Operations Planner and Scheduler (OPS) was created. Successfully optimizing operations from an energy and water quality/supply perspective requires both software tools and the human resources necessary to develop a Daily Operating Plan (DOP). For most water utilities, CSU included, this also requires an organizational change to accept the concept of a centralized point for coordination of system operations and maintenance. Additional system operations technical resources are also required to develop the DOP. System operators are too involved with daily operations and typically lack the technical skills to successfully produce and coordinate an optimized operating plan.

CSU's experience in creating, designing and implementing the OPS function is presented in this chapter of the report. The methodology employed by CSU to create the OPS may be used by any water utility considering a systems perspective for operations planning and scheduling.

SYSTEM OPERATIONS PLANNING AND SCHEDULING

Prior to the EWQMS project, CSU coordinated operations between water supply, treatment, and collection/distribution on an informal basis. The system worked reasonably well over the years, but lacked the structure required for optimized operations.

It was proposed to CSU management that the water utility adopt a systems approach to operations similar to the electric utility – that is, strategic operational planning and scheduling of Raw Water Production, Water Transmission, Water Treatment, Distribution, Wastewater Collection, and Wastewater Treatment in a manner similar to planning Fuel Supply, Generation, Transmission, and Distribution operations in the electric utility. The electric utility has over 30-years of experience in structured system operations planning yielding high reliability at the lowest possible cost to the electric customer.

The focal point for coordinated water utility operations is the OPS. An OPS may be a single person or a group of individuals and supporting software. In either case, the OPS's mission is to develop a daily operating plan. The daily operating plan contains the operating schedule for the water and wastewater systems. The schedule shows which pumps must run, how PRVs are set, how much surface water is used, etc., for each time period in the day. The plan will be used to align and coordinate water and wastewater system operations and maintenance. The daily plan is developed to ensure that the water system meets customers' demand for water while complying with water quality requirements, controlling energy cost, and maximizing hydro-generation utilization.

The primary role of the OPS is water system coordination to enable execution of the DOP. Optimization of water system operations requires accurate forecasting of water consumption and scheduling production, raw water transmission, pumping, treatment, and distribution system valves to minimize cost and maximize product quality and reliability. The OPS depends on the predicted availability of water system apparatus (pumps, valves, production capacity, etc.) to meet the operating plan. Formal operating procedures are required to remove or constrain equipment operation or change the topology of the system by operating valves. The OPS adjusts the operating plan as necessary to respond to unscheduled events (e.g., equipment failure, line breaks). Success in meeting the operating plan is dependent on the operational discipline of the organization to adhere to formal procedures and not make system changes without consulting with the OPS.

IMPLEMENTATION OF THE OPS FUNCTION

CSU took a systematic approach to creating the OPS function. This section of the report defines the methodology used by CSU to create the OPS.

Establish an EWQMS Project

The first step in creating the OPS was to establish an EWQMS project. Based on review of previous AwwaRF/EPRI work of the EWQMS concept and perceived opportunities in Colorado Springs, CSU management created an EWQMS project and assigned a project manager. Capital funding was established for the planning phase of the EWQMS and a successful proposal was written for collaborative funding from AwwaRF to share the project experience industry-wide.

Create and Fill an OPS Position

CSU management made a strong commitment to the EWQMS concept by creating a formal project. It was also a belief that the OPS role should be assumed by an experienced technical resource at the beginning of the project to provide continuity between design and implementation.

A needs assessment job description was written for an OPS position (Appendix B). The position was filled prior to formal initialization of the project.

Form an EWQMS Implementation Team

Implementation of the EWQMS at CSU was approached as a business process optimization project in alignment with other reengineering projects under way at the utility. An Implementation Team (I-Team) was recruited representing all the functional operational areas of the CSU water and wastewater system. Members on the I-Team also included representatives from water quality, information technology, planning, construction, and the electric utility.

The responsibilities of the EWQMS I-Team are summarized in the following Purpose Statement:

> "The Energy & Water Quality Management System (EWQMS) I-Team will help to complete the implementation of an EWQMS at CSU, including the formation of an Operations Planning/Scheduling function that will support CSU's values regarding the community, our customers, and our employees. The team will work to promote collaboration within CSU to ensure that operations are communicated and coordinated through a formal planning and scheduling process resulting in improved operational efficiency and effectiveness."

The EWQMS I-Team held structured meetings and recruited content experts as necessary to define the functionality of operations planning and scheduling.

The Mission Statement created by the EWQMS I-Team provided a strong foundation for building the OPS concept at CSU:

The EWQMS I-Team will complete the planning, implementing and tracking phases of the project by:

I-Team Formation:

- Developing an effective EWQMS I-Team
- Communicating EWQMS to the organization
- Obtaining management & organizational support for EWQMS

Planning/Design:

- Supporting the modeling/design of the OPS function
- Supporting the development/utilization of a Water Operations DOP
- Communicating the OPS function and the DOP to the organization

Implementation:

- Promoting cultural change necessary to implement EWQMS
- Encouraging and supporting EWQMS principles and OPS function

Tracking:

- Developing metrics, monitoring, improving and documenting EWQMS processes
- Verifying effective implementation of EWQMS

The I-Team then created a set of Guiding Principles upon which they would operate. Implementation of the EWQMS and OPS represented a significant operational and cultural change for CSU, and these Guiding Principles provided direction for the I-Team in their interactions with internal and external stakeholders.

1. CSU and CSU-Water Resources Department's (WRD) vision, mission, and strategic plan must drive the development of an OPS function.
2. As the EWQMS OPS function is developed, this team will make many decisions. Those decisions will be made with the goal of improving CSU's competitive position and in returning value to CSU's citizen owners. Individual department positions will not be used to override decisions that are best for CSU as a whole.
3. Communication with WRD and CSU employees and stakeholders will be up-front and comprehensive on all aspects of the development of the OPS function.
4. The development of the OPS function will be accomplished with a high degree of collaboration with those responsible for water consumption forecasting, water and wastewater quality, water rights maintenance, energy cost scheduling, the WRD Program Driven Maintenance Project, and the WRD/CSU Information Technology group. The EWQMS I-Team will consist of subject matter experts (SMEs) from each one of these disciplines from within WRD and/or from CSU as appropriate.
5. The final EWQMS implementation plans and recommendations will result in considerable cultural changes regarding work practices within WRD and other CSU departments. In order to assure the success of the project, assessing organizational readiness and planning for these changes will be critical. The I-Team will utilize the

CSU Organizational Development Consultant (ODC) to evaluate its implementation and communication plans before implementation of the OPS process and function.

6. The Primary Project Sponsor, the Executive Sponsor, and the Water Resources Department Director must approve the OPS process and overall project plan before actual implementation begins.
7. It is recognized that the EWQMS project has several external stakeholders due to the uniqueness and scope of its work. Some of these external stakeholders are the American Water Works Association Research Foundation (AwwaRF), EMA, Inc., and the EWQMS Project Advisory Committee. Each of these external groups has separate time frames and project deliverables that are dependent upon the work of the I-Team. The EWQMS I-Team will do its best to help these external groups achieve their project goals in a timely manner. However, the I-Team will not compromise these guiding principles in order to comply with the needs of the project's external stakeholders.
8. It is recognized that the EWQMS Project is a WRD Optimization Project approved by the WRD Strategic Leadership Team (SLT) and the WRD Director. Therefore, the WRD Optimization Project Manager will be principally responsible for integrating the work of the EWQMS Project Manager and all I-Team members into the overall WRD Optimization process.

Define the EWQMS Domain

The EWQMS project touches many organizational groups within and external to CSU. For this reason, it is important to define the EWQMS domain. "What" it is, and "what" it is not. Figure 3.1 illustrates what functions are included in the EWQMS and those functions outside the EWQMS domain.

Create an EWQMS Vision

An EWQMS Vision Statement was created by the I-Team so all team members and other CSU employees would have a common understanding of the future goal or direction of the project.

> "EWQMS applies technology appropriately to establish agreed-upon operational processes for optimizing energy and water quality management. These processes provide for a scheduled work plan which is derived through collaboration with all CSU stakeholders using standard procedures, and is readily available to all operations personnel in the form of a daily operating plan, regularly updated and communicated electronically. By enabling collaboration in preparation of the operating plan, EWQMS facilitates interdepartmental alignment and coordination by viewing CSU as a single entity, with all benefits accruing to CSU as a whole. This cooperation of partners who eschew organizational rivalries in favor of the greater good of CSU results in improved efficiency and effectiveness, as well as enhanced quality of service and system reliability. Resources are conserved and protected, as well. CSU's customers benefit from the deferral of capital improvements and reduced O&M cost in the form of lower rates and a

dependable, high-quality water supply. As a consequence, they view CSU as a standard-bearer in the utility industry and as a clear choice as their provider of utility services."

Define Strategies to Bridge From Today to the Vision

The next level of detail required to construct the EWQMS was the definition of strategies to bridge from the current state of operations to the future state of operations illustrated by the Vision. The following Strategies were grouped by technology, training, communication, organization, performance and tracking, implementation, and business practice.

Technology: Provide reliable technologies that support all aspects of the EWQMS.

- Consumption Forecast Program provides daily water consumption for each service level. System simulator predicts water system performance based on forecasted consumption.
- Modified Water Information System (WIS) database provides information that will be passed to/from the EWQMS. The RMS2000 will be used for clearance request work orders and to communicate existing and planed equipment outages. AI SCADA provides real-time and historical water system information. EWQMS will send a pump and valve schedule for control by AI SCADA. Water and Wastewater Treatment Plants will transfer information to/from Plant Control Systems (PCS) to the EWQMS. The Geographic Information System Water Quality Analyzer continuously receives water quality information, compares this information to regulations and CSU performance criteria and adjusts operating constraints and/or alarms to the OPS. Upgrade the Evapo-Transpiration (ET) weather stations for more accurate water consumption forecasting. Integrate data in a data repository.
- Develop Pump Optimizer to minimize the cost of pumping. A Hydro Optimizer maximizes the benefit of hydroelectric generation for CSU Electric.

Training: Training will be required for EWQMS development and effective change.

- Train all stakeholders on relevant aspects of the EWQMS. Provide information to all CSU employees on EWQMS project philosophies.
- Reinforce training with sound EWQMS operating and implementation plan.

Communication: Communicate information about EWQMS and communicate between systems for collaboration.

- Develop web-based communications applications for EWQMS.
- Initiate weekly planning meetings with all stakeholders.
- Initiate a clearance process for the Water System.
- Develop and work a communication plan.

Organization: Organizational change management and Electric Department participation is necessary for EWQMS development.

- Enlist and maintain support of CSU leadership.

- Integrate relevant aspects of Program Driven Maintenance (PDM) Project with EWQMS.
- Define and assign responsibility and authority for all aspects of OPS processes.
- Eliminate current "silo" mentality of divisions and departments related to utility system operations.
- Develop and institutionalize OPS functions and roles.

Performance and Tracking: Performance metrics will allow near real-time tracking of relevant trends and show benefits of the EWQMS.

- Develop performance metrics for the EWQMS.
- Define CSU performance criteria for water system operations.

Implementation: Implement the EWQMS CSU-wide.

- Implement EWQMS in progressive steps, starting small and building out.
- Change our business practices to support EWQMS.
- OPS is the single point of contact with energy providers for operations/energy cost scheduling in the water/wastewater system.

Construct an Asset Model

The I-Team next constructed a water/wastewater system Asset Model to understand the utility from a systems perspective of the utility. The location and type of energy production and consumption assets located in the water system. Figure 3.2 illustrates a section of the Asset Model. Note that the type of pump station motors, instrumentation and control equipment are identified in the Asset Model.

The Asset Model was structured after the water utility business model defined in the AwwaRF *The Utility Business Architecture: Designing for Change* report (Brueck, Rettie, and Rousso 1998). A high level abstraction of the CSU water/wastewater business model is shown in Figure 3.3. The EWQMS domain includes the optimum operation of the water functional areas shown in Figure 3.3.

Develop an EWQMS Model

The I-Team developed the EWQMS Model previously shown in Figure 1.2 with knowledge gained from the previous steps. The EWQMS Model illustrates the functional components and inputs/outputs (I/O) of an EWQMS system. The model is used extensively to develop the functional requirements for software and OPS processes.

Identify the EWQMS Processes

An EWQMS Process Model was created by the I-Team to identify the business practices associated with the EWQMS. The EWQMS Process Model is shown in Figure 3.4.

The nine processes in the model define the work necessary to plan and schedule water system operations at CSU. Each of these processes was modeled and used to define organization and technology requirements for both the EWQMS and OPS.

Model the EWQMS Processes

Each EWQMS process was modeled by the I-Team with assistance from subject matter experts. The processes were called "Use Cases" because the team modeled how the organization would "use" an EWQMS. Note that a strict adherence to use-case modeling for software development was not followed. The team was more focused on the organizational dynamics of an EWQMS than rigorous software design.

The nine use-cases are illustrated in Figures 3.5 through 3.13. The use-cases in the figures show each task in the process, the individuals or "actors" who perform each task, time duration of the task, and decision blocks with a percentage of time for each branch from the block. A definition of each use-case is shown in the figures along with the glossary of terms.

The I-Team also estimated the number of occurrences, or instances, of each use-case. Based on this information, the EWQMS Process Model was simulated for one year. This simulation provided a first-cut estimate for the time required for the OPS function. The simulation identified over 7,000 hours per year for OPS time. This result identified the requirement to share OPS responsibilities between the designated OPS, System Operators, and other CSU staff.

Institutionalize the OPS

Modeling the use-cases provided an understanding and a future vision of how the EWQMS would work at CSU. They provide a blueprint of the new operational practices and software requirements.

The next step, institutionalizing the OPS, requires management buy-in, consistent communication, and strength of conviction that the EWQMS will significantly improve utility system operations. It is a difficult cultural change to accept the notion of centralized system planning and scheduling. The previously defined work in this chapter of the report provides the structure and necessary planning to minimize the problems associated with the organizational change.

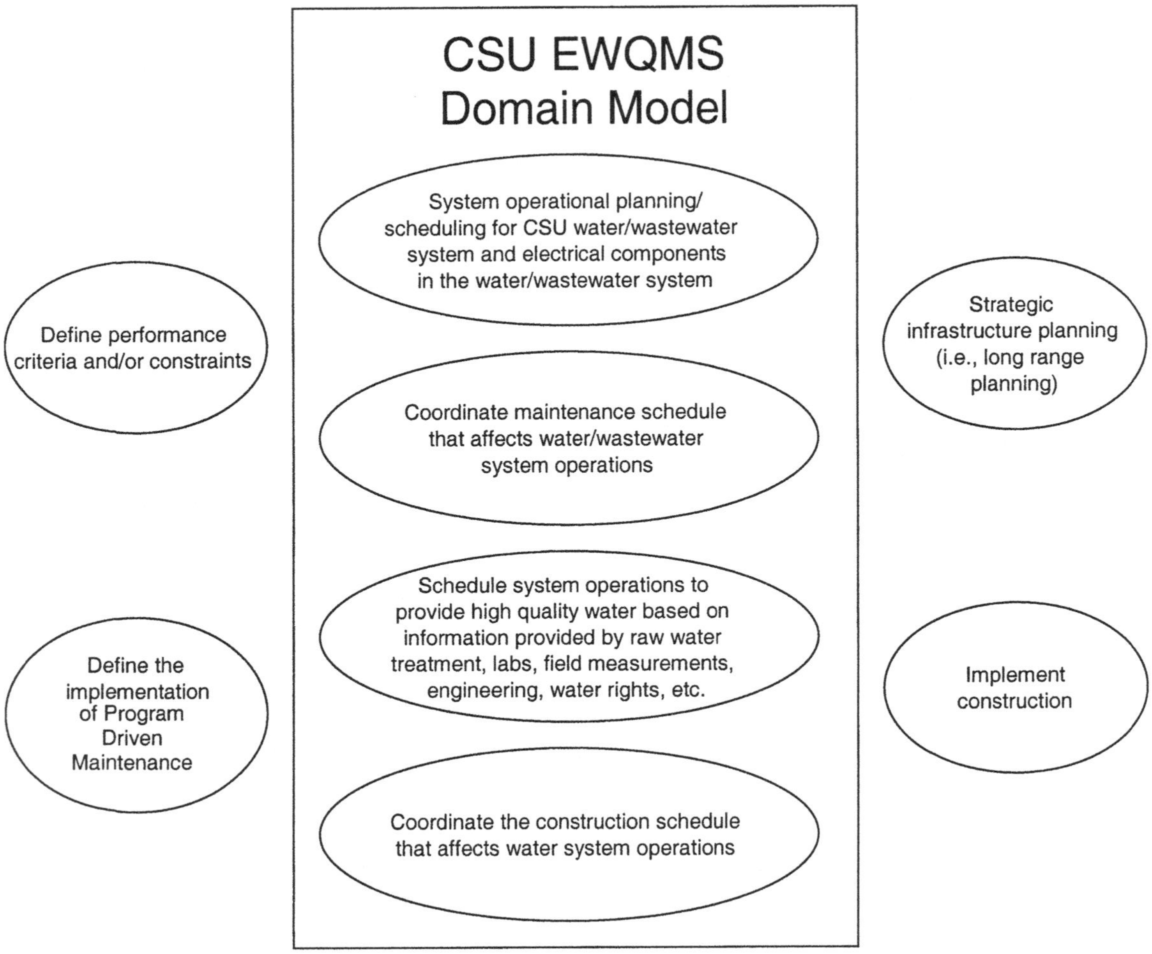

Figure 3.1 CSU EWQMS Domain Model

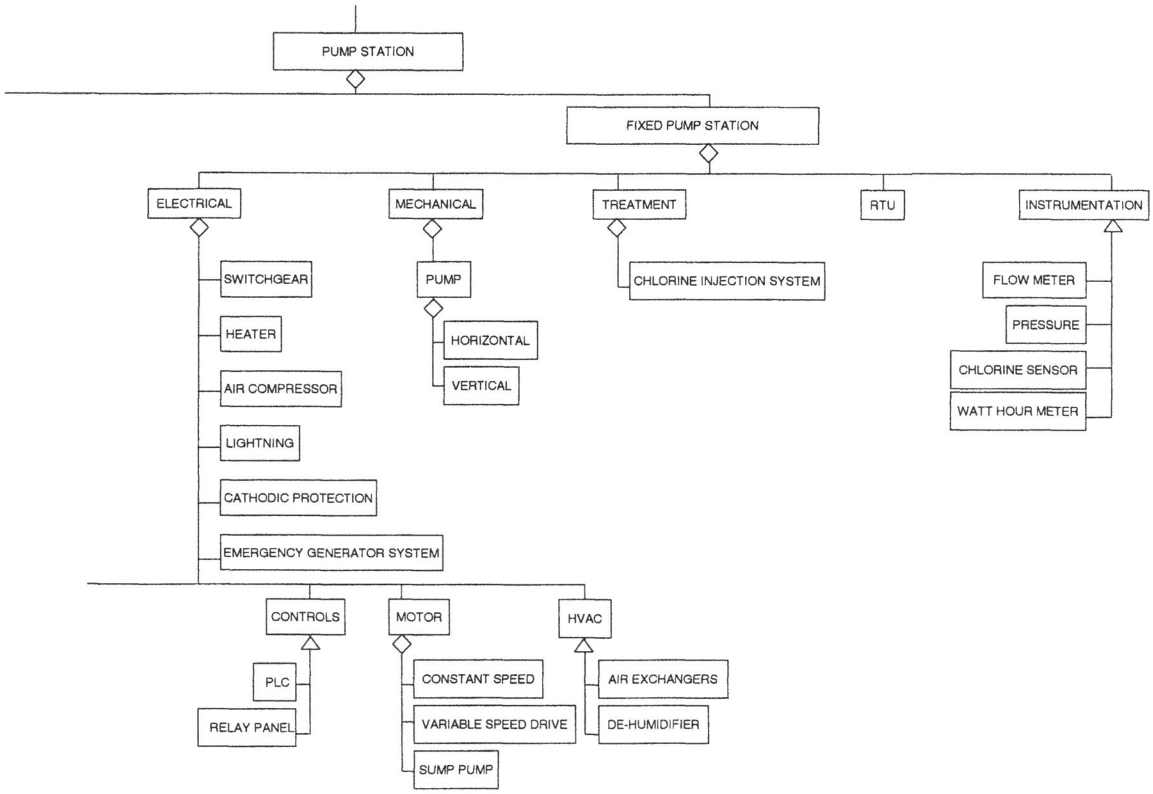

Figure 3.2 CSU Asset Model – sectional view

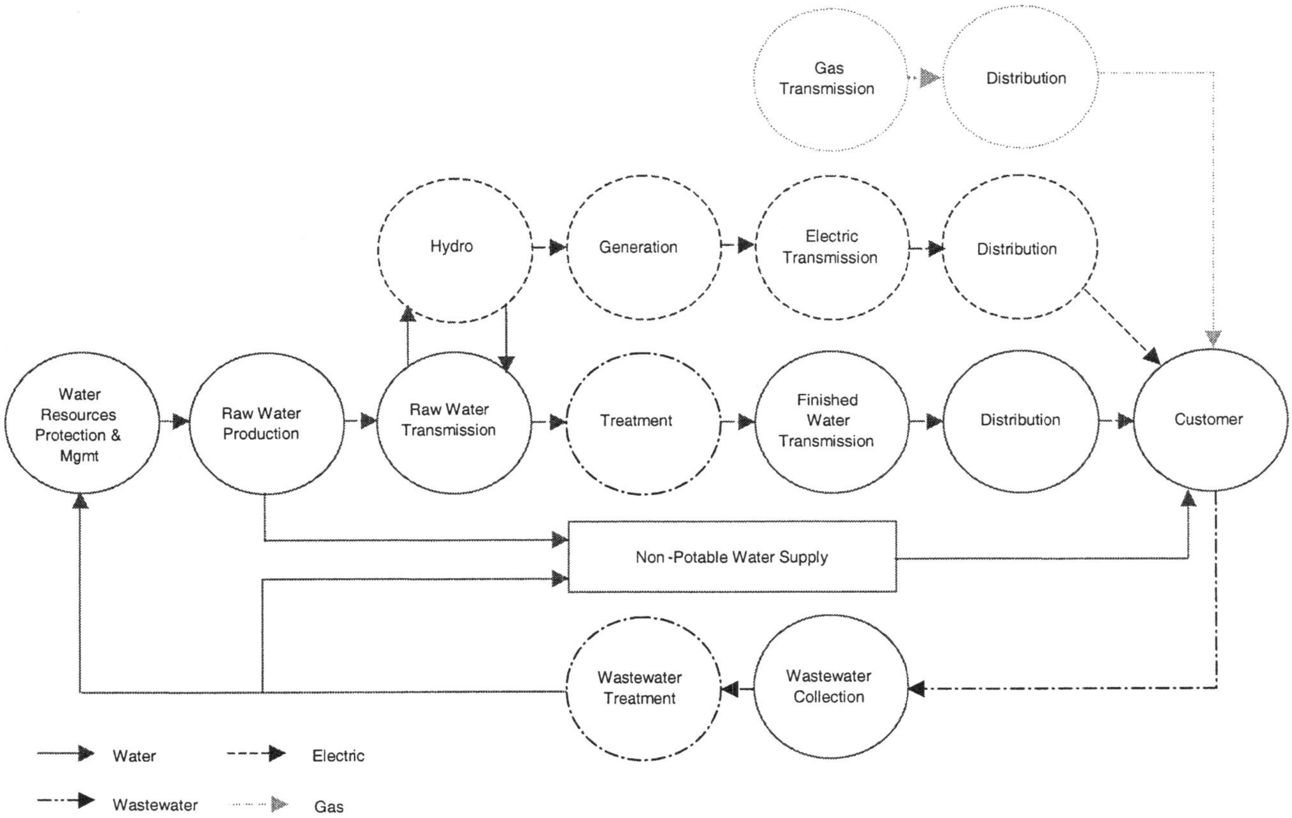

Figure 3.3 CSU Operations Business Model

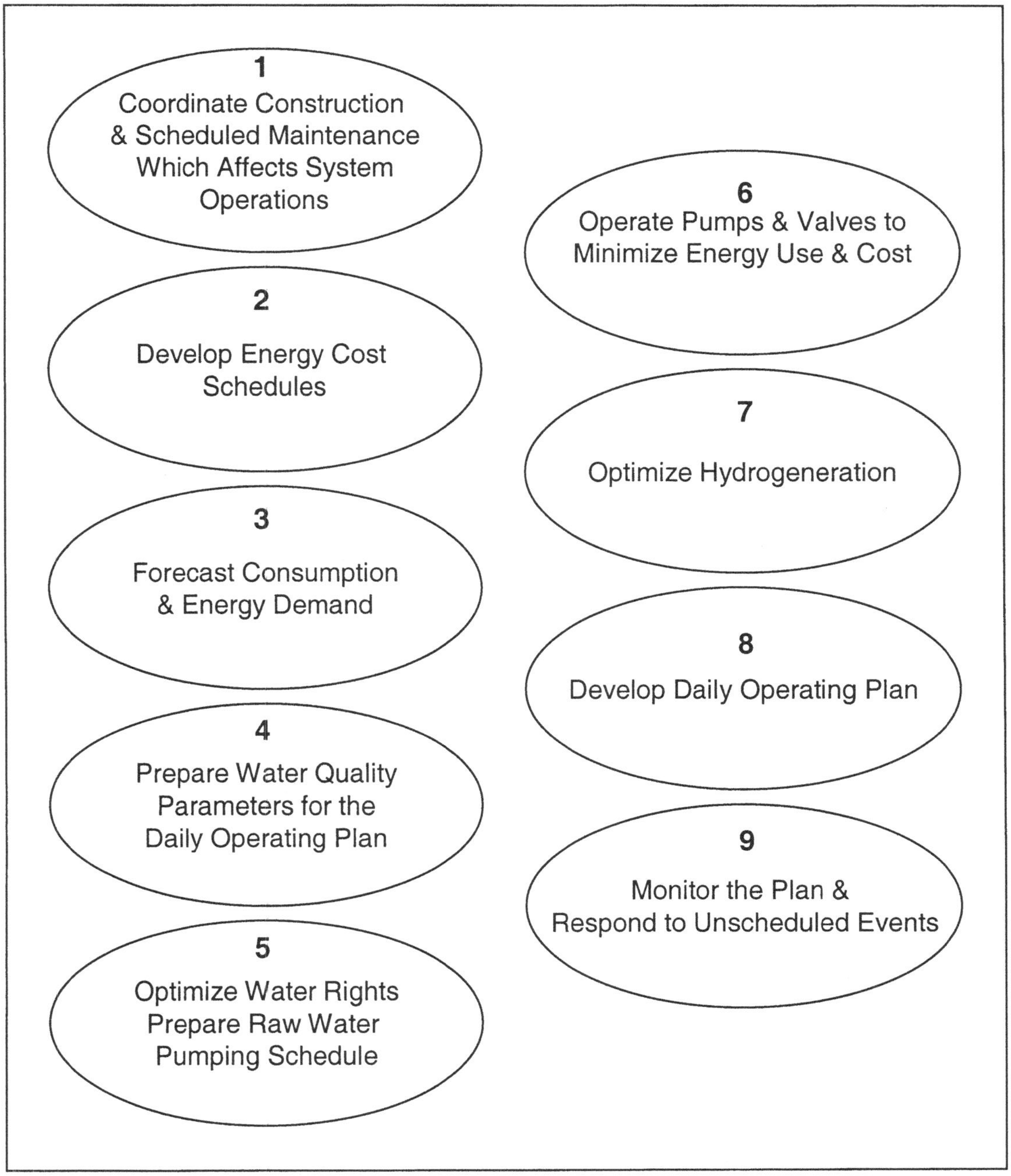

Figure 3.4 EWQMS processes included in the Process Model

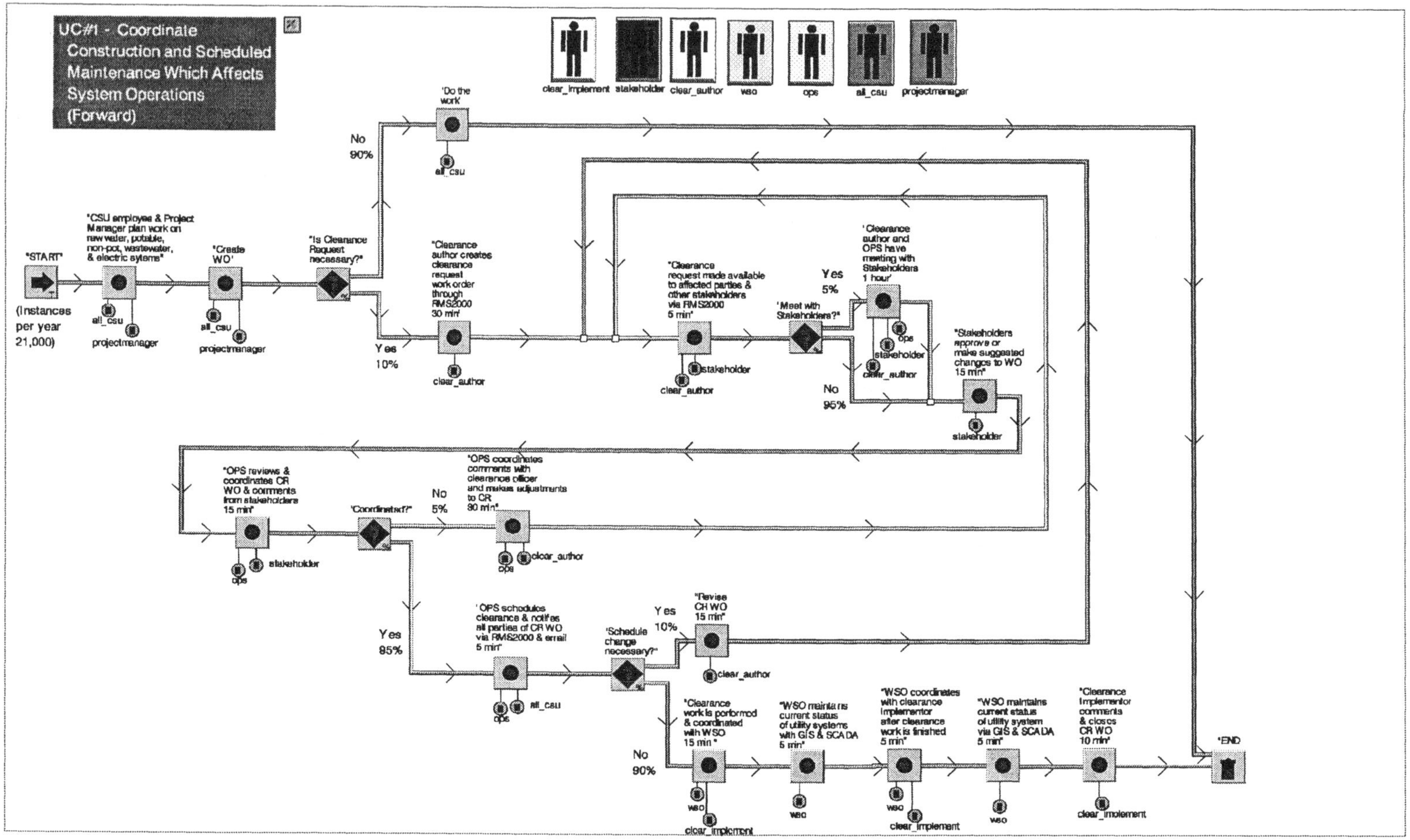

<table>
<tr>
<td>

Definition:

Collaborate to identify, prioritize, schedule, implement and plan for construction and maintenance activities for Raw Water, Potable, Non-potable, Wastewater & Electric systems to maximize systems reliability, minimize costs and ensure water quality as it impacts system operations.

</td>
<td>

Glossary:

1. all_csu = Any person with responsibility to EWQMS
2. clear_author = Clearance Author
3. wso = Water System Operator
4. ops = OP/S, Operations Planner Scheduler
5. clear_implement = Clearance Implementor
6. GIS = Geographical Information System
7. SCADA = Supervisory Control and Data Acquisition
8. CR = Clearance Request
9. WO = Workorder

</td>
</tr>
</table>

Figure 3.5 Use Case #1 – coordinate construction and scheduled maintenance which affects system operations

(An 11 x 17 version of this figure is available at the back of the report)

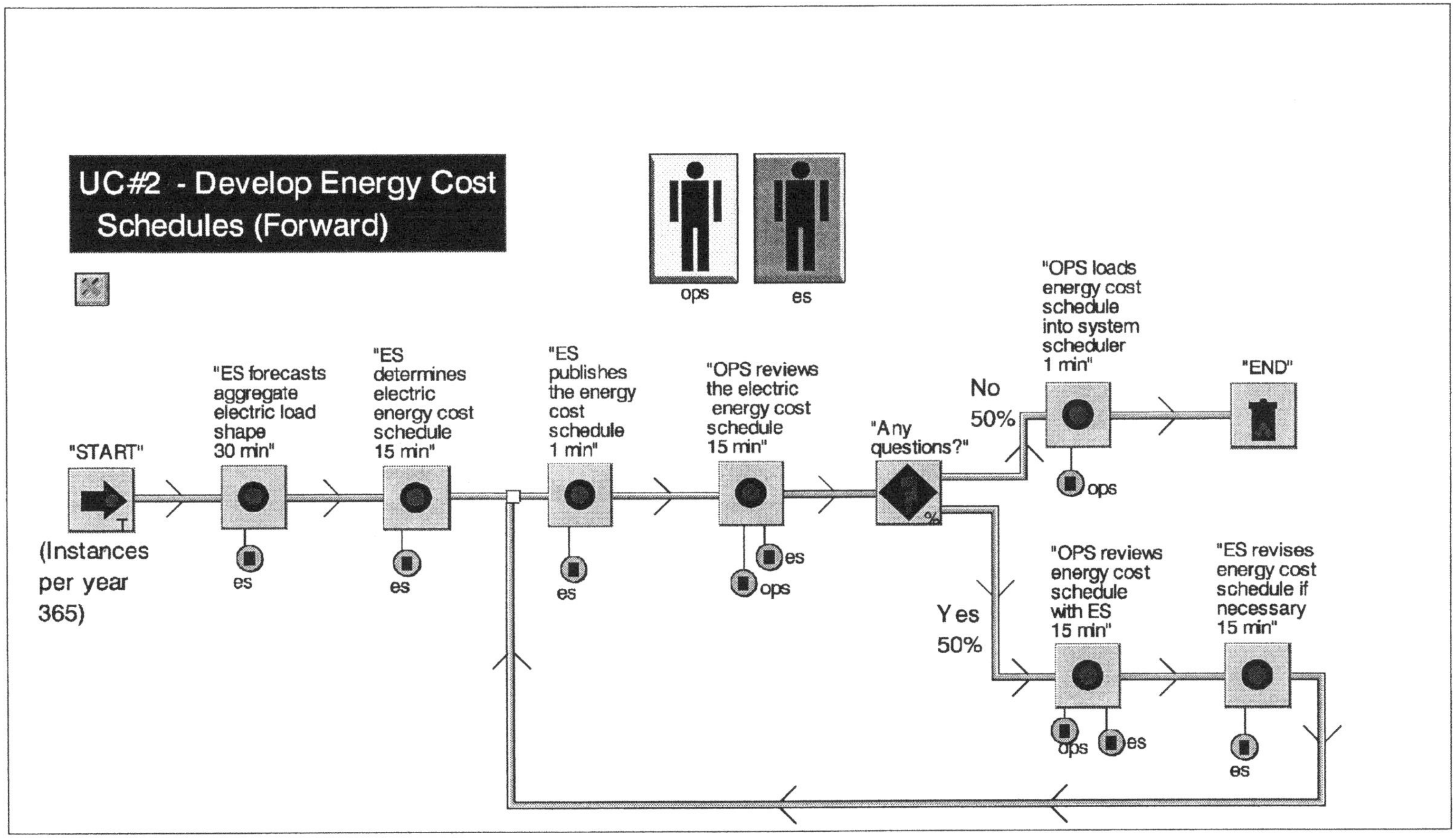

Definition:

CSU Energy Supply prepares an energy cost schedule on a daily basis for use by the EWQMS.

Glossary:

1. ops = OP/S, Operations Planner Scheduler
2. es = Energy Staff

Figure 3.6 Use Case #2 – develop energy cost schedules

(An 11 x 17 version of this figure is available at the back of the report)

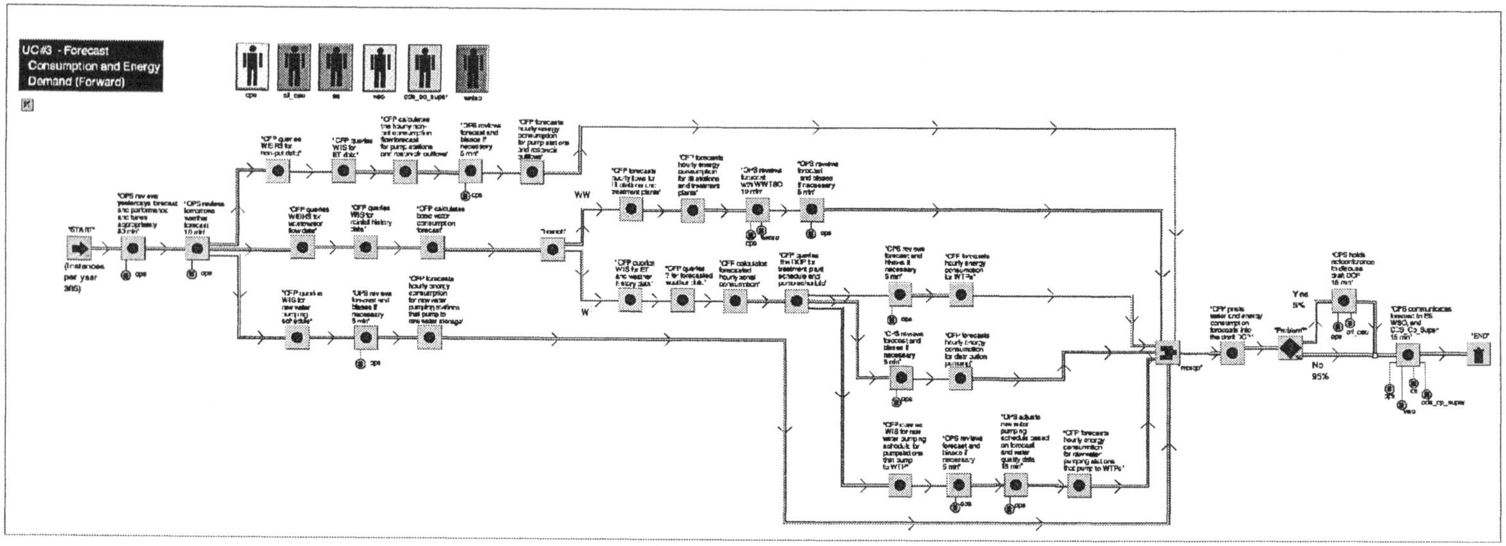

Definition:

The OP/S prepares a water consumption forecast on an hourly basis with zonal resolution. The OP/S collaborates with Electric in the preparation of their electric system consumption forecast.

The OP/S prepares a wastewater flow forecast on an hourly basis with a lift station, and plant resolution.

Glossary:

1. es = Energy Staff
2. wso = Water System Operator
3. cds_op_super = Collection Distribution Services Operations Supervisor
4. DOP = Daily Operating Plan
5. WIS = Water InformationSystem
6. CFP = Consumption Forecast Program
7. WEIRS = Wastewater Environmental Information Retrieval system
8. ET = Evapo-Transpiration
9. Non-Pot = includes raw water and re-use
10. WWTSO = Wastewater Treatment System Operator
11. WTP = Water Treatment Plant
12. ops = OP/S Operations Planner Scheduler
13. all_csu = Any person with responsibility to EWQMS

Figure 3.7 Use Case #3 – forecast consumption and energy demand

(An 11 x 17 version of this figure is available at the back of the report)

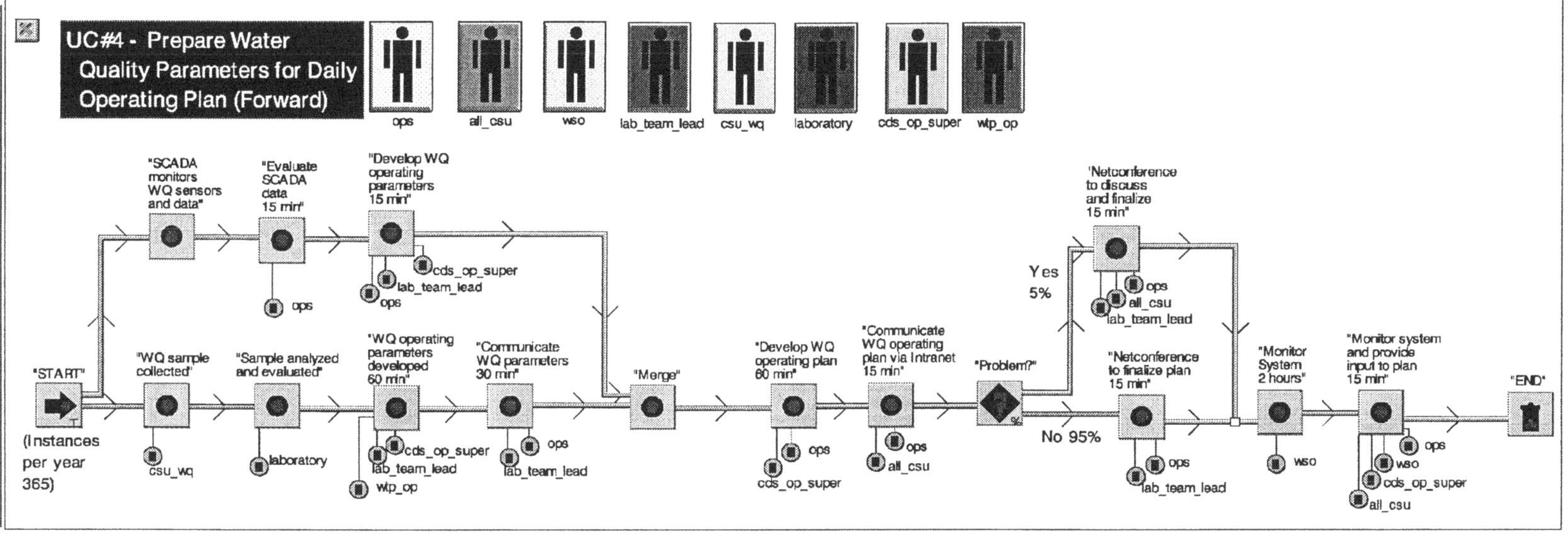

Definition:

Evaluate water quality information such that daily operating parameters are prepared as a basis for the system operating plan. This includes raw water, non-potable, potable, and wastewater.

Glossary:

1. all_csu = Any person with responsibility to EWQMS
2. wso = Water System Operator
3. ops = OP/S, Operations Planner Scheduler
4. csu_wq = Any person in CSU collecting samples
5. cds_op_super = Collection Distribution Services Operations Supervisor
6. lab_team_lead = Laboratory Team Leader
7. WQ = Water Quality
8. Laboratory = Lab Team
9. WTP_Op = Water Treatment Plant Operator

Figure 3.8 Use Case #4 – prepare water quality parameters for daily operating plan

(An 11 x 17 version of this figure is available at the back of the report)

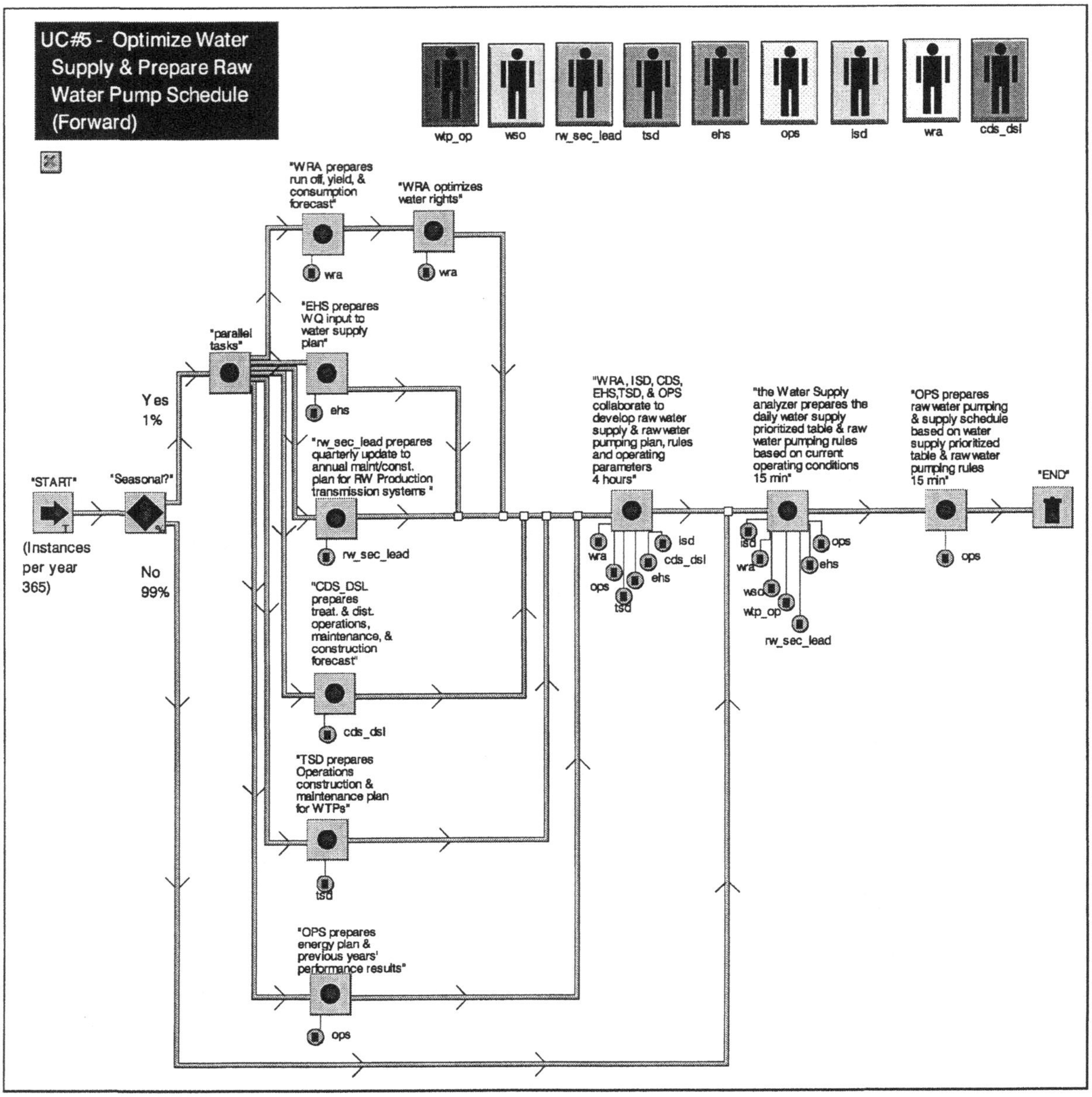

Definition:

The OP/S prepares the Raw Water Supply and Raw Water Pumping Schedule using rules from the Water Supply Analyzer.

Glossary:

1. wra = Water Rights Administrator
2. ehs = Environmental Health & Safety
3. RWSL or rw_sec_lead = Raw Water Section Leader
4. ops = OP/S, Operations Planner Scheduler
5. cds_dsl = CDS Control & Distribution Section Leader
6. treat. = Treatment
7. dist. = Distribution
8. WQ = Water Quality
9. const. = Construction
10. WTP = Water Treatment Plant
11. tsd = Treatment Service Division
12. rw_tsd = Raw Water Treatment Service Division
13. isd = Information Services Division

Figure 3.9 Use Case #5 – optimize water supply and prepare raw water pump schedule

(An 11 x 17 version of this figure is available at the back of the report)

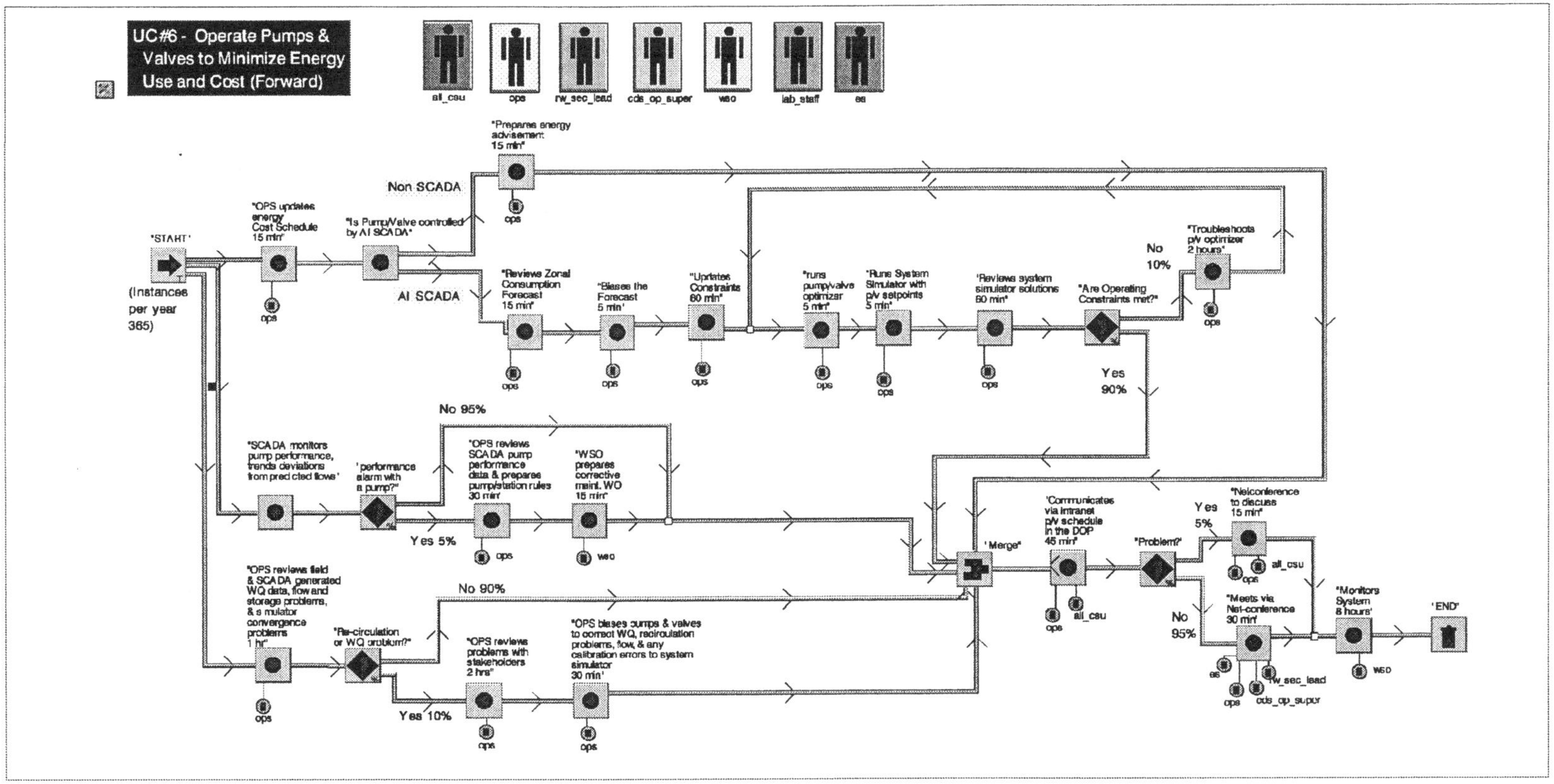

Definition:

OP/S prepares daily/weekly pump/valve setpoint tables that control pumps/valves based on energy usage and cost, operating constraints/requirements, and water quality/supply in the distribution, transmission, raw water transmission, raw water production, wells, and non-potable systems. The OP/S will not prepare pump/valve schedules for water treatment plants and the wastewater system. The OP/S will provide an energy advisement for water treatment plants and the wastewater system.

Glossary:

1. all_csu = Any person with responsibility to EWQMS
2. wso = Water System Operator
3. ops = OP/S, Operations Planner Scheduler
4. rw_sec_lead = Raw Water Section Leader
5. cds_op_super = Collection Distribution Services Operations Supervisor
6. lab_staff = Laboratory Staff
7. DOP = Daily Operating Plan
8. p/v = pump/valve
9. WO = Work Order
10. es = Energy Staff

Figure 3.10 Use Case #6 – operate pumps and valves to minimize energy use and cost

(An 11 x 17 version of this figure is available at the back of the report)

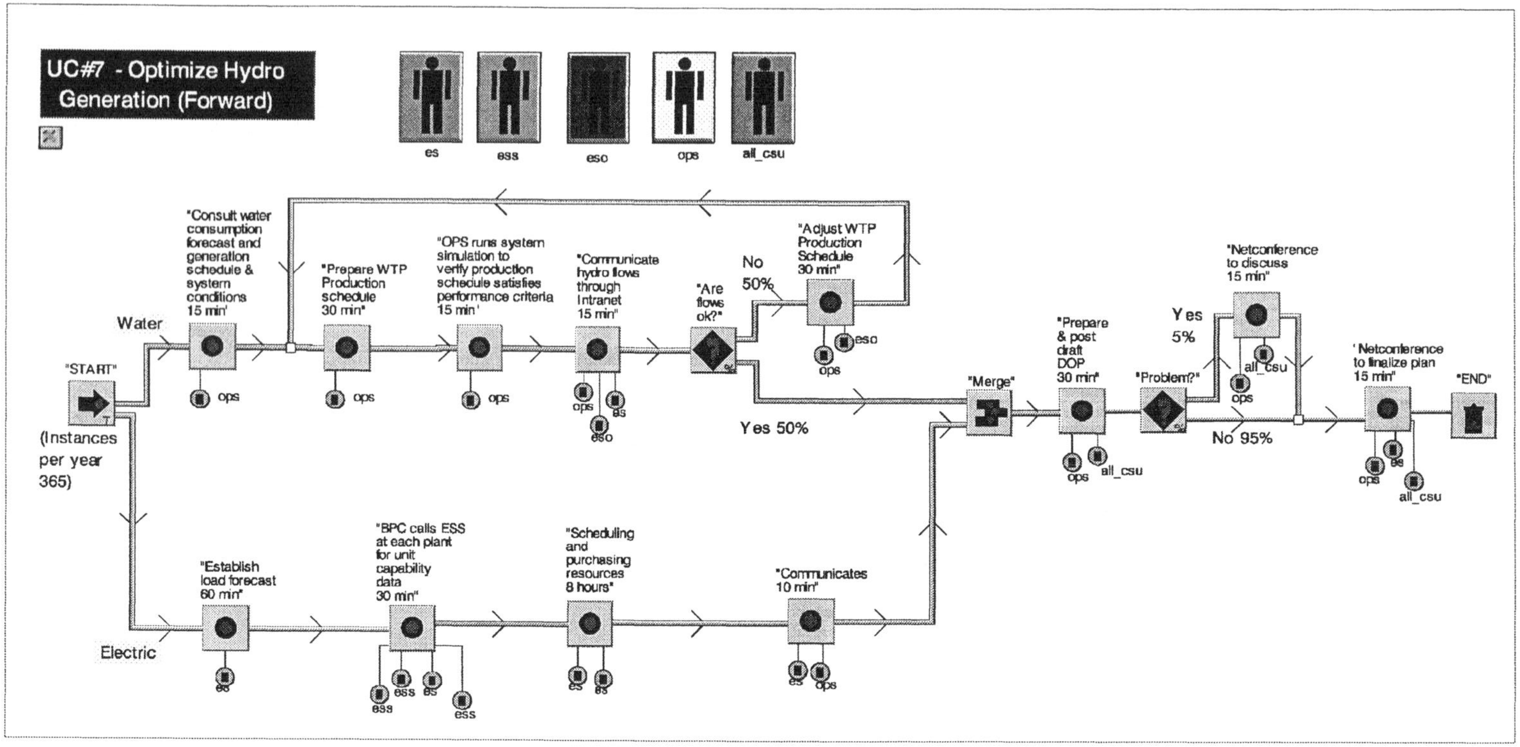

Definition:

Electric and Water collaborate to optimize the operation of CSU Hydro within the constraints of water demand, spinning reserve, and system limitations.

Glossary:

1. all_csu = Any person with responsibility to EWQMS
2. ops = OP/S, Operations Planner Scheduler
3. eso = Electric System Operator
4. es = Energy Staff
5. ess = Electric Shift Supervisor
6. WTP = Water Treatment Plant
7. DOP = Daily Operating Plan
8. WRD = Water Resource Department

Figure 3.11 Use Case #7 – optimize hydroelectric generation

(An 11 x 17 version of this figure is available at the back of the report)

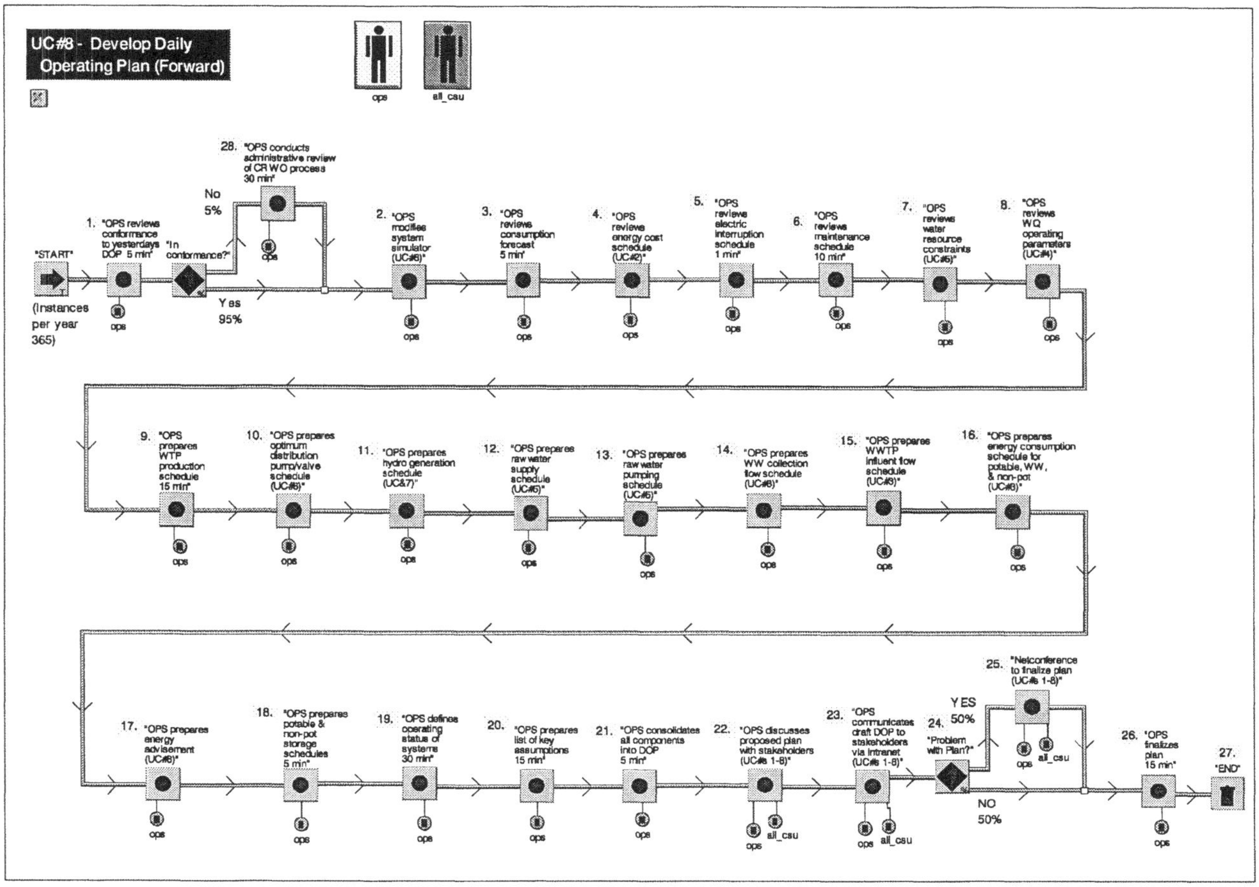

<u>Definition:</u>

OPS prepares daily operating plan including: distribution pump/valve schedule; hydro generation schedule; raw water pumping schedule; water treatment plant production schedule; wastewater collection flow schedule; energy consumption schedule for water, wastewater, & non-potable systems; water quality operations parameters, water resource constraints e.g., water rights, priorities, & capacities; projected operating status of the water, wastewater, & non-potable systems; potable & non-potable water storage schedule; list of key assumptions, raw water supply schedule, and energy advisement.

The Operating Plan is based on a potable & non-potable and consumption forecast and construction & maintenance schedules. The plan is prepared to minimize O&M costs, maintain water quality, and provide high system reliability for potable, non-potable, and wastewater customers.

<u>Glossary:</u>

1. all_csu = Any person with responsibility to EWQMS
2. WTP = Water Treatment Plant
3. ops = OP/S, Operations Planner Scheduler
4. DOP = Daily Operating Plan
5. WWTP = Wastewater Treatment Plant

Figure 3.12 Use Case #8 – develop daily operating plan
(An 11 x 17 version of this figure is available at the back of the report)

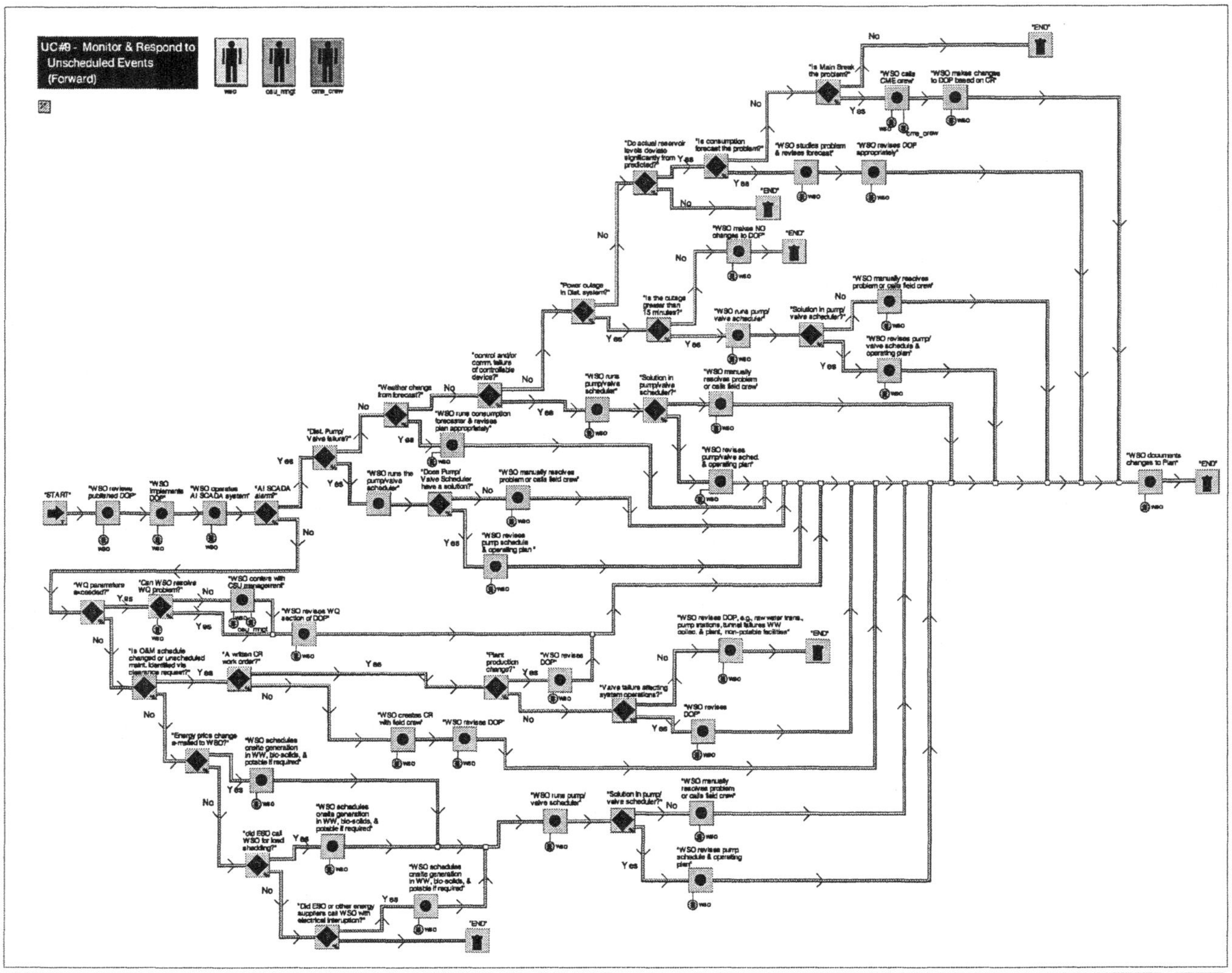

Definition:

WSO implements the Plan (DOP) and reviews information from a number of sources to determine compliance with the Plan. The WSO is the point of contact for any unscheduled event which affects the operation of potable, non-potable, raw water, and wastewater systems. The WSO revises the DOP based on the unscheduled event. The WSO documents every deviation to the DOP.

Glossary:

1. EHS = Environmental Health Services
2. CME = Construction & Maintenance Equipment
3. WTP = Water Treatment Plant
4. DOP = Daily Operating Plan
5. wso = Water System Operator
6. Dist. = Distribution
7. Comm. = Communication
8. WQ = Water Quality
9. csu_mngt = CSU Management
10. O&M = Operations & Management
11. Maint. = Maintenance
12. WW = Wastewater
13. eso = Electric System Operator
14. CR = Clearance Request
15. WO = Work Order
16. Collec. = Collection
17. Cme_crew = Construction & Maintenance Equipment Crew

Figure 3.13 Use Case # 9 – monitor and respond to unscheduled events (An 11 x 17 version of this figure is available at the back of the report)

CHAPTER 4
FUNCTIONAL SPECIFICATIONS FOR EWQMS SOFTWARE

CONTEXT OF SPECIFICATIONS

The EWQMS Functional Specifications were written from two perspectives:

1. To define the specific requirements for an EWQMS at CSU
2. To provide a template for EWQMS specifications that can be used by other water utilities

The specifications were written to procure, design, code, test, and put software into operation that performs the functions of an EWQMS at CSU.

Chapter 1 of this report defined the high level functionality of the EWQMS, which reflects an ultimate EWQMS implementation. System implementation takes place in two phases. In the first phase, software requirements were defined to meet the following criteria:

1. Can be operational within six months after purchase
2. Has a positive ROI within twelve months after placed into operation
3. Is appropriate technology for the OPS

The purpose of these criteria is to minimize complexity and cost of systems selected for implementation.

The second phase will include those components of the EWQMS not implemented during Phase 1. The requirements will be defined after the first phase is implemented.

The following defines the components of the Phase 1 implementation of the EWQMS:

1. Consumption Forecaster Program
2. Economic Pump/Valve Controller
3. System Scheduler consisting of:
   a. Water System Simulator
   b. Optimum Pump/Valve Scheduler
4. Water Quality Analyzer Process for Water Resource Scheduling
5. Water Source Analyzer
6. Daily Operating Plan
7. Daily Operating Plan Netconference
8. Construction/Maintenance Clearance Process

SPECIFICATION DESCRIPTIONS

The components of the EWQMS were illustrated in Chapter 1, Figure 1.2. This chapter describes the specifications for the first EWQMS phase.

The Consumption Forecaster program will predict hourly consumption for each service level in the distribution system. Phase 1 Specifications for the Consumption Forecaster are found in Appendix C.

The AI SCADA System will monitor and control the water distribution, wastewater collection, and non-potable systems. Fixed automatic setpoints and time-based economic setpoint tables will be used by the SCADA system to control pumps and control valves in the water system. The Economic Pump/Valve Controller provides the interface between the AI SCADA system and the System Scheduler software. Specifications for the Economic Pump/Valve Controller are located in Appendix D.

The System Scheduler will use a mass-balance water system simulator, pump schedule, treatment plant schedule, and water source schedule to minimize the cost of pumping and optimize hydro-electric generation. Appendix E contains the specifications for the Water System Simulator, and the Optimum Pump Scheduler is located in Appendix F.

The Water Quality Analyzer will focus on a surrogate water quality indicator in the distribution system – water age. The Water Quality Analyzer specifications are found in Appendix G.

The Water Source Analyzer reviews forecasted consumption data and develops an optimum schedule for the multiple water sources based on water rights, historical flows, pumping cost, storage value, local source value, hydro-generation potential, and water quality. Specifications for the Water Source Analyzer are included in Appendix H.

The Daily Operating Plan (DOP) uses information from other EWQMS applications to communicate the plan for optimized operations. Appendix I contains the specifications for the DOP.

The DOP is communicated via CSU's Intranet to interested stakeholders. Specifications for the DOP Netconference application are found in Appendix J.

The Construction/Maintenance Clearance Process is required to maintain the current and predicted state of the water system for simulation and optimization. Appendix K includes specifications for the Construction/Maintenance Clearance Process.

CHAPTER 5
PROCUREMENT, DEVELOPMENT, AND INSTALLATION OF EWQMS SOFTWARE

The process used by CSU to procure, develop and install EWQMS software is described in this chapter. The time to complete this phase of the project was approximately 12 months. Testing and calibration of the EWQMS software required additional time.

PROCUREMENT

During the period that the CSU/Consultant team was working on EWQMS software specifications, a parallel review of potential software products was completed. No commercially available software products met the requirements of the EWQMS specifications. However, the team used off-the-shelf products as foundations upon which the applications were built.

- Construction/Maintenance Clearance Process – Lotus Notes
- Consumption Forecaster – Microsoft Excel
- Daily Operating Plan – Microsoft Excel
- Daily Operating Plan Netconference – Lotus Notes
- Economic Pump/Valve Controller – Metso Automation OASyS
- Optimum Pump Scheduler – Microsoft C++
- Water Quality Analyzer – Incorporated with Water System Simulator (Gensym G2)
- Water Source Analyzer – Microsoft Excel and Evolver
- Water System Simulator – Gensym G2

CSU used existing AI SCADA and Consultant contractors to procure software and services to implement the EWQMS to minimize time and cost. The utility was experiencing severe budget restrictions, and funding on the EWQMS was cut from previous capital plans. A design goal to implement simple, low cost, and high return applications was a necessity because of these funding limitations.

DEVELOPMENT

Development of the EWQMS benefited from a teaming approach between CSU, consultants, and suppliers. Frequent review meetings were held via teleconference and web conference. A web-based team room was used to post design memorandum documents and comments from team members. Face-to-face meetings were the exception.

All participants in the software development phase understood the nature of this project. Software was being created that had no precedence. For this reason, it was necessary to be patient and work together as a team to solve problems.

Functional specifications were modified and enhanced as necessary to reflect the details of implementation. The specifications contained in this document are second generation and improved from the initially prepared version.

Development of operations optimization software is highly dependent on the utility topology and composition of water system components. Although standard approaches can be used, actual implementation depends on the unique nature of each water system. The software development team relied heavily on the knowledge of CSU operators, managers, and planners to build the models, which approximated actual operations of the water system.

Data availability was a problem during development. To build, maintain, and calibrate operations models requires archived data, which may be difficult to obtain or non-existent. To resolve this problem, CSU used surrogate data (datapoint calculated from sensor data) and new instrumentation needs were defined.

INSTALLATION

EWQMS software was installed on a designated server on the CSU AI SCADA system. Displays and reports are available to the OPS and System Operators on the SCADA Local Area Network (LAN). After the software was installed, operations staff were trained on functionality of the software. Two levels of training were required – user and administrator.

User training was focused on system operators. They must have a thorough understanding of the EWQMS software functionality and design basis. The system operator plays an important role in the tuning, calibration, and operations of the EWQMS software. It is important for them to be comfortable with software because it will automate many of their current functions. They must know how to monitor progress of the daily operating plan and intervene as necessary to maintain water system reliability if conditions change as the result of unscheduled events. System operators also provide important feedback to the OPS for software calibration.

Administrator training is required for the OPS and other technical staff. This training is focused on the internals of the EWQMS software. The OPS must know how to make changes to the logic, graphics, and database to support water system modifications.

TESTING AND CALIBRATION

Testing and calibration of the EWQMS software is a long and complicated process. Full functionality of the EWQMS software is heavily dependent on the ability to accurately portray the actions of the water system. Bad SCADA data, network configuration problems, and pumps operating off their curves are examples of contributors to uncalibrated models. Directly affecting this accuracy are system attributes such as: reservoir capacity, pump flow rates, and transfer percentages. To refine these attributes to match historical trends, the models are first run using historical SCADA data. Results from the models are compared to actual operating data. By comparing the simulated trends to historical trends, attributes are adjusted to match actual trend performance.

CSU's hydraulic model is used to configure and calibrate the water system simulator and optimum pump scheduler. The water system simulator is a rules-based high-level representation of the water system whose topology is configured and operating rules developed, using knowledge gained and results produced from the detailed hydraulic model.

CSU will not place the EWQMS software into operation until it is thoroughly tested and calibrated. A large population of historical test days over various consumption levels is required. After the operations staff is satisfied with the results of historical testing, the system will be cut over to optimized system control in a sequential fashion by areas in the water system. First an isolated section of the water system will be tested on-line. After this area operates successfully, other areas will be added, step-by-step, until the entire system is under EWQMS control.

CHAPTER 6
FINDINGS AND RECOMMENDATIONS

This chapter documents the results of the CSU EWQMS project and describes the lessons learned and observations from the team members about their experiences. It also contains recommendations for any water utility considering a future EWQMS project.

RESULTS

The process of developing and implementing an EWQMS at CSU was enlightening. The potential benefits of planning and scheduling daily operations are significant and many. The EWQMS Implementation Team began the project with the expectation that CSU would primarily benefit by reducing energy expenditures, with the possibility of improving water quality, or at the very least not degrading it. The reality, after arduous reengineering of processes and development of functional specifications for software applications, is that the utility has much to gain by aligning and ordering the operational processes in such a way that information about the short-term schedules of all operational functions are available to all operational units and systems before and during a work period. It took considerable effort to document and quantify many of the potential benefits of EWQMS due to a shortage of available supporting data.

One of the discoveries the team made early on in the process analysis phase of the project was that CSU needed a process module which would perform the task of analyzing and optimizing the sources of water to be used to meet the demands of the Daily Operating Plan, considering cost, water rights, and storage issues. We added a "Water Source Analyzer" to the set of process modules comprising CSU's EWQMS. This was in addition to the AwwaRF EWQMS Reference Model.

Another decision was that incremental development of the Water System Simulator and Optimum Pump Scheduler portions of the Operations Planner/Scheduler function was needed, since the Extended Period Simulation (EPS) capabilities of the available distribution system model were not sufficiently well developed to enable its use directly as a planning and scheduling tool. A simplified simulator/scheduler using a software tool with a more flexible front end and a less arcane instruction set was specified. At a future time this decision may be revisited, in the event that the EPS model has matured to make it more of an operational tool useful for short-term decisions pertinent to daily scheduling.

From the beginning of the work to develop the Maintenance Scheduler, it was anticipated that the existing CSU RMS2000 work management system would provide the basis for this application, perhaps through the development of a specialized work order that enable additional approval procedures that would enable clearance of requested work that would affect system operations. As it turned out it was much simpler to design a clearance process using a simple application in Lotus Notes, which is in use throughout CSU. Those work orders describing work which will potentially affect system operations are designed to be entered into the application for approval or "clearance" prior to scheduling them through RMS2000. These clearance requests are then routed to all stakeholders via Lotus Notes for their tacit approval before granting the clearance.

As strenuous and time consuming as are process reengineering and preparation of software specifications, the organizational implications of changing the operational culture of a water utility are significant and require much time and effort to achieve. Change is difficult to accomplish in the best of circumstances, but especially so in water utility operations, which are traditionally reactive in nature. Those who perform the work of system control are often very accomplished at "seat-of-the-pants" operational processes and may be slow to acknowledge that what works for them is not necessarily best for the enterprise as a whole. Changing their mandate from monitoring in real-time and reacting as necessary to operating to conform to the Daily Operating Plan can create difficulties for some operators and their managers, especially in view of the fact that they often carried out their old mandate very successfully. It can take them some time to see the benefits of operational planning and scheduling which takes into account the needs of the entire utility process "value chain" (UBA).

FINDINGS

Implementing an EWQMS is a large undertaking involving many players coordinating efforts to develop individual applications that work together to provide information to the client. The CSU project was comprised of people from a number of organizations located throughout the country, which complicated project management and communications aspects of the project. The EWQMS itself is comprised of numerous applications working together to provide information. These applications were developed simultaneously requiring significant communication and coordination.

Some of the challenges that were encountered as the project progressed are listed below:

A. Team Communication: Difficulties working with a large collaborative, integrated team located throughout the country consisting of:
   1. Client (CSU)
   2. Consultant
   3. Outside Contractors
   4. AwwaRF (providing oversight of project)

B. Project Management: Due to the number of applications being developed simultaneously, standard project management techniques need to be in place

C. Budget and Scope of Work: Difficulties with budget and project scope being reduced mid-way through the project

D. Integration with Client Organization: Integration with the management of the client organization throughout the project is critical

E. Data Collection and Verification: Due to the fact that the data comes from many different sources, conflicts in data arose

F. Technology Issues:
   1. "Shrink-wrap" solutions do not exist at this point and optimization applications in water system operation are in their infancy
   2. Creation of the mass-balance model was the largest technical challenge

G. Management of Expectations and Results:
   1. Expectations of optimization must be realistic and clearly defined
   2. Results need to be produced early on in the project and discussed at length

RECOMMENDATIONS

Team Communication

Budget sufficient time for face-to-face meetings, especially in the beginning phase of the project. It is important for all of the team members to have a common design framework in front of them and to fully understand what is expected of them from the beginning. Technology assists diverse teams to communicate, (i.e., the Team Room web page, web conference calls, etc.) but it cannot replace the necessity of face-to-face meetings. Once the project is clearly defined and understood, and quality controls are in place, everyone is then responsible for completing their given tasks, utilizing the technology to enable communications. Further face-to-face meetings should occur periodically, specifically at the final review and testing phases.

Project Management

Standard project management techniques, such as the development of a critical path for the timing of application development, must be in place. It is important for each team member to meet critical path deadlines because project tasks and activities are dependent on one another. The start/end times should be established up front and a project manager involved to keep all participants on track.

Budget and Scope of Work

A dedicated project team must have adequate funding and time to complete the tasks. Consistency in utility management is required to implement the organizational changes necessitated by implementation of an EWQMS.

Integration with Client Organization

It is important that the EWQMS team have support and good communication with the leadership of its client organization. Solicit active and ongoing support from those at the highest levels of the organization, and continuously publish their endorsements widely to ensure the unstinting sponsorship of the project. Implementation of EWQMS represents a fundamental change in the way a utility conducts its business. Progressive and knowledgeable leadership is required to see the implementation through to the end. If this commitment has not been obtained, or if the perseverance of leadership is questionable, it may be advisable to delay further efforts until follow-through is assured. The EWQMS team should be proactive in working with the IT leadership in an organization so all parties are working with them to work together and adhering to standards whenever possible.

Data Collection and Verification

Although the specifications for CSU's data requirements were well developed, the need to develop and organize a data collection team from the onset of the project is essential to orchestrate the collection of data. Collection should be a separate task with its own funding and staff who are responsible for the collection and verification of data. This team should resolve any conflicting data early in the project.

Technology Issues

"Shrink-wrap" solutions have not been developed and optimization applications in water system operation are in their infancy. Each application must be developed and tailored to the specific water system being optimized.

Creation of the mass-balance model was the project's largest technical challenge. There were many discussions via web conference to finalize the model. This process might have been easier if the group had met face-to-face to create this base model.

Management of Expectations and Results

Expectations of optimization need to be realistic and clearly defined. The dialogue between the Water and Electric functional areas of CSU was extremely worthwhile in terms of stakeholders reaching consensus on the meaning of optimization. The client must be able to identify simple patterns and rules corresponding to optimum operation so the results that are obtained can be verified.

Results, even for a partial system, and even if sub-optimal, need to be produced early and discussed at length. It is critical that *end-users* evaluate results at a detailed, technical level. Initially, this evaluation itself may not be well defined. It is very important to present results in a format compatible with the way the user thinks about optimizing the system.

Implementation

The following are recommendations for a water utility considering an EWQMS:

1. Assess the opportunity and quantify benefits of an EWQMS – the benefits provide the business case to implement an EWQMS
2. Provide a clean and concise report to management on the opportunities and cost of an EWQMS
3. Develop a vision for future operations using an EWQMS, including a definition of optimum operations
4. Develop an EWQMS model based on the AwwaRF reference model
5. Define and model the energy and water quality/supply management processes associated with operations planning and scheduling
6. Implement the organizational process change first, then implement the EWQMS software technology
7. Develop functional software specifications based on the benefits analysis and operational modeling
8. Implement software solutions incrementally – start first with simple solutions, especially those which make use of technology already owned, and build out with more complexity if there is a commensurate return on investment

9. Determine if existing SCADA data is sufficient to meet the requirements for operational models and provide the resources necessary to install and calibrate field instrumentation where required
10. Allow time to tune and calibrate EWQMS models based on the complexity of the water system and the availability and accuracy of SCADA data
11. Collaborate with the water utility's energy provider to develop energy cost schedules, as well as to identify opportunities for mutual benefit and shared cost savings
12. EWQMS is an integrated system solution and should be implemented with all components to attain the full benefits of the system.

APPENDIX A
WATER QUALITY DATA

Plots of 1998-1999 Simulated Water Age and Chlorine Residual at Distribution System Sampling Stations.

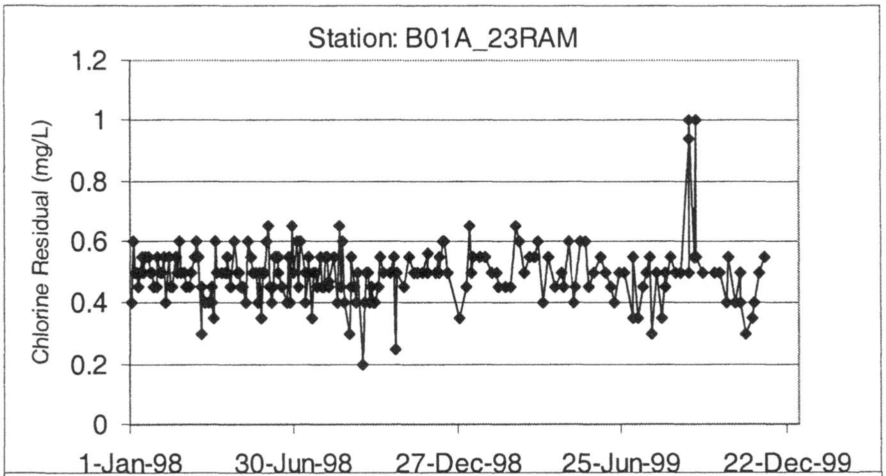

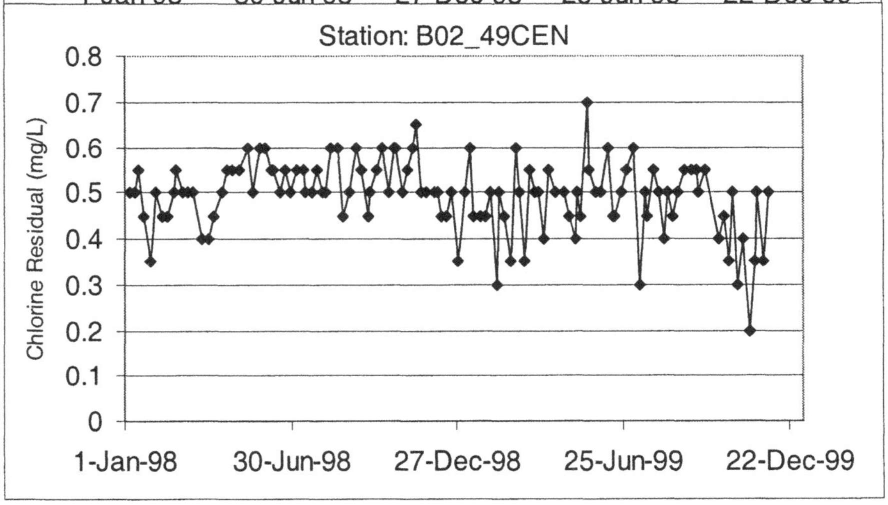

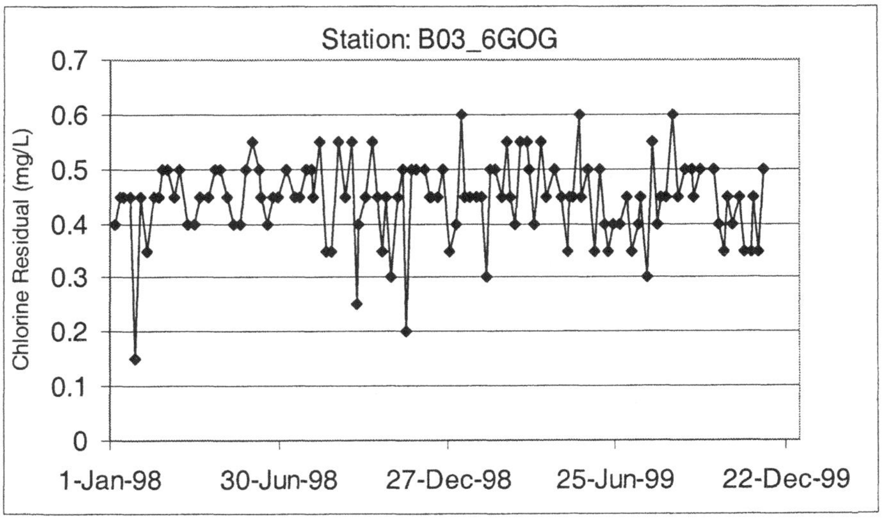

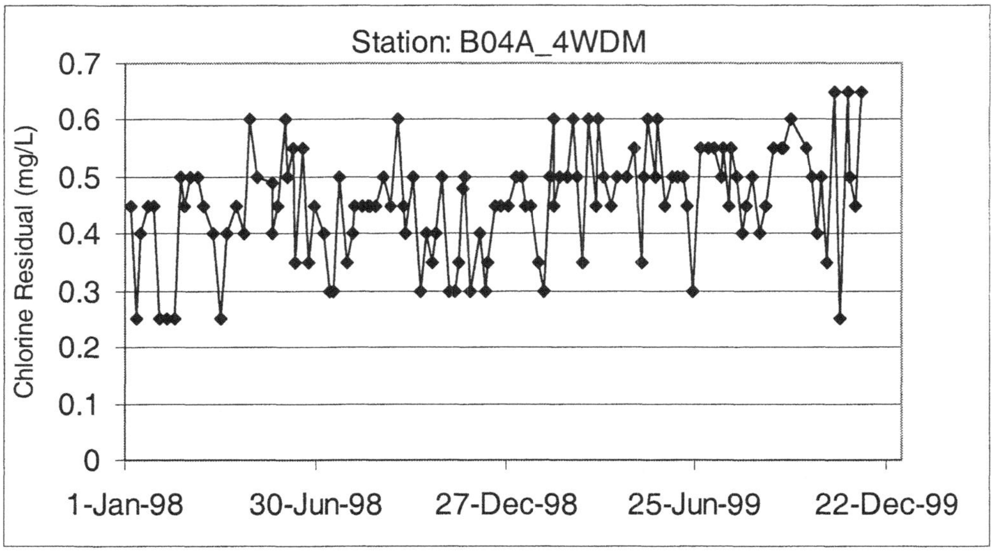
Station: B04A_4WDM
0.7
0.6
0.5
0.4
0.3
0.2
0.1
0
Chlorine Residual (mg/L)
1-Jan-98
30-Jun-98
27-Dec-98
25-Jun-99
22-Dec-99

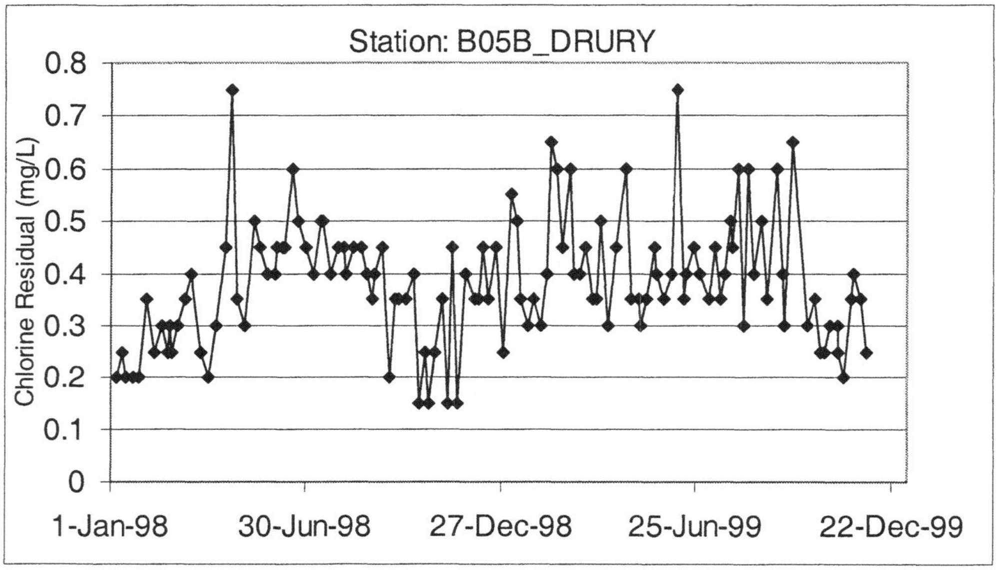
Station: B05B_DRURY
0.8
0.7
0.6
0.5
0.4
0.3
0.2
0.1
0
Chlorine Residual (mg/L)
1-Jan-98
30-Jun-98
27-Dec-98
25-Jun-99
22-Dec-99

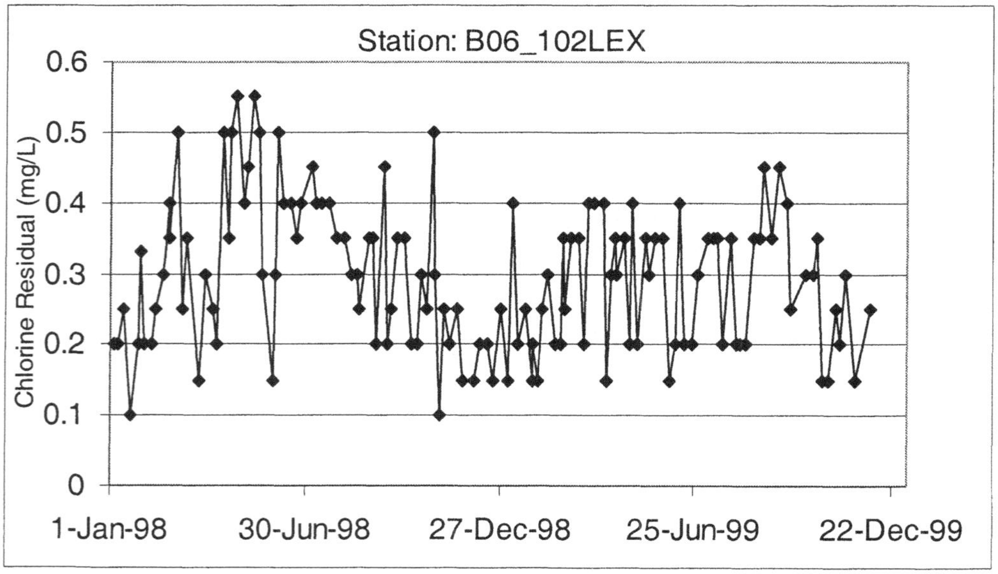
Station: B06_102LEX
0.6
0.5
0.4
0.3
0.2
0.1
0
Chlorine Residual (mg/L)
1-Jan-98
30-Jun-98
27-Dec-98
25-Jun-99
22-Dec-99

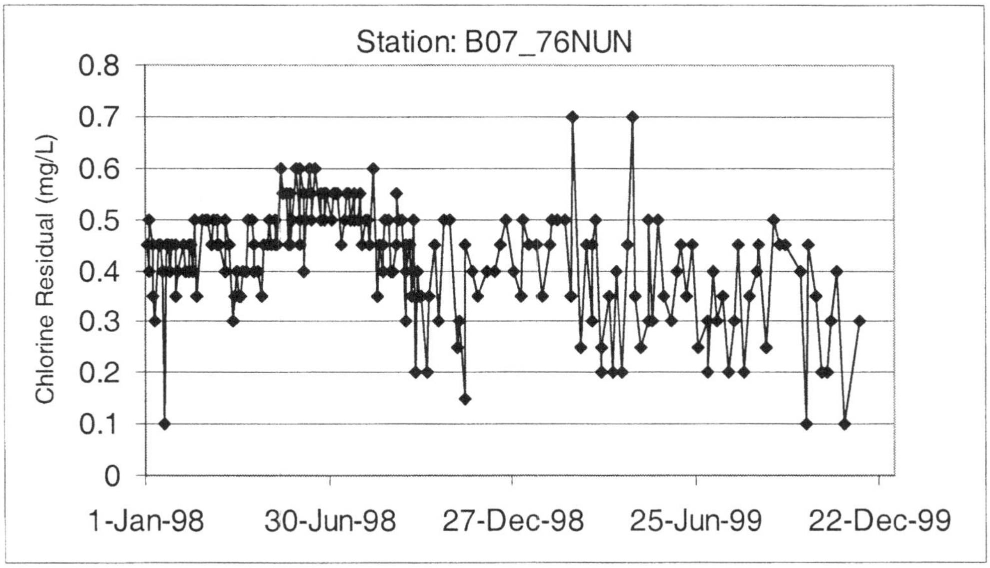
Station: B07_76NUN
Chlorine Residual (mg/L)
0.8
0.7
0.6
0.5
0.4
0.3
0.2
0.1
0
1-Jan-98
30-Jun-98
27-Dec-98
25-Jun-99
22-Dec-99

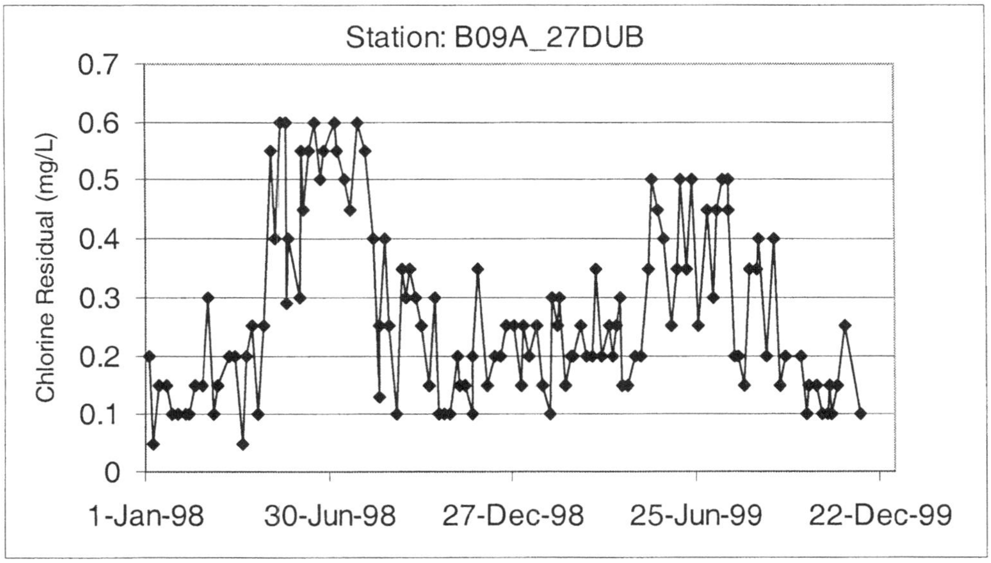
Station: B09A_27DUB
Chlorine Residual (mg/L)
0.7
0.6
0.5
0.4
0.3
0.2
0.1
0
1-Jan-98
30-Jun-98
27-Dec-98
25-Jun-99
22-Dec-99

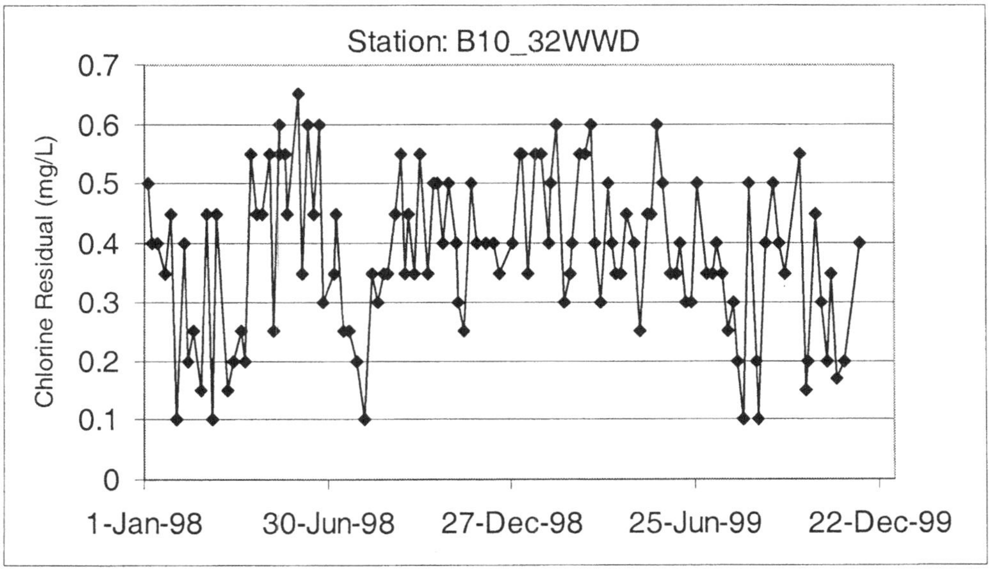
Station: B10_32WWD
Chlorine Residual (mg/L)
0.7
0.6
0.5
0.4
0.3
0.2
0.1
0
1-Jan-98
30-Jun-98
27-Dec-98
25-Jun-99
22-Dec-99

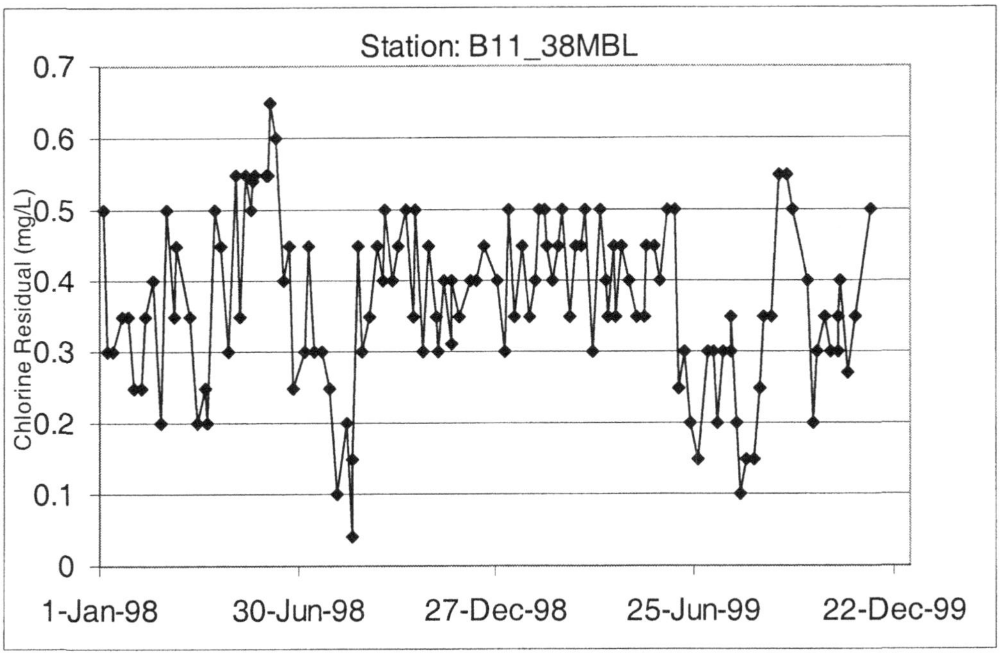
Station: B11_38MBL
Chlorine Residual (mg/L)
0.7
0.6
0.5
0.4
0.3
0.2
0.1
0
1-Jan-98
30-Jun-98
27-Dec-98
25-Jun-99
22-Dec-99

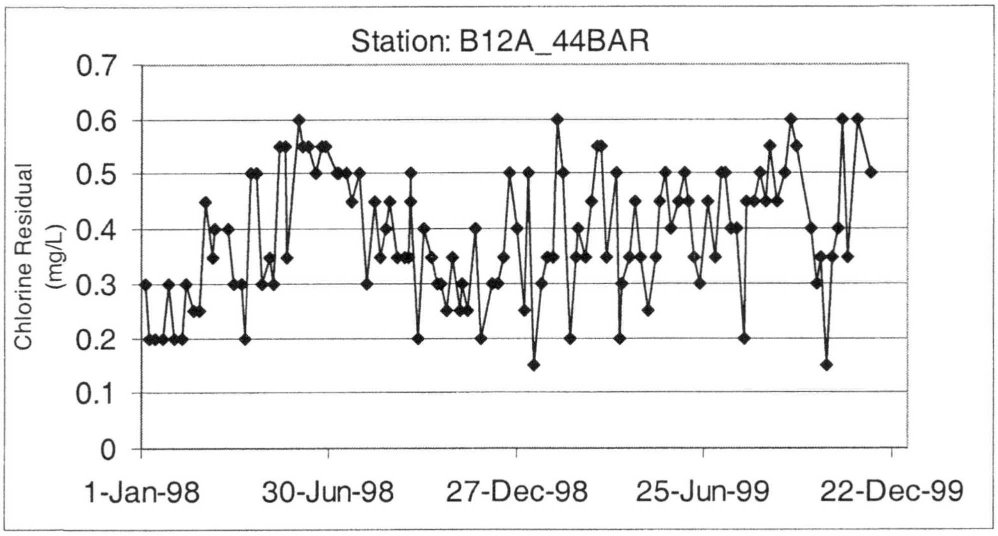
Station: B12A_44BAR
Chlorine Residual (mg/L)
0.7
0.6
0.5
0.4
0.3
0.2
0.1
0
1-Jan-98
30-Jun-98
27-Dec-98
25-Jun-99
22-Dec-99

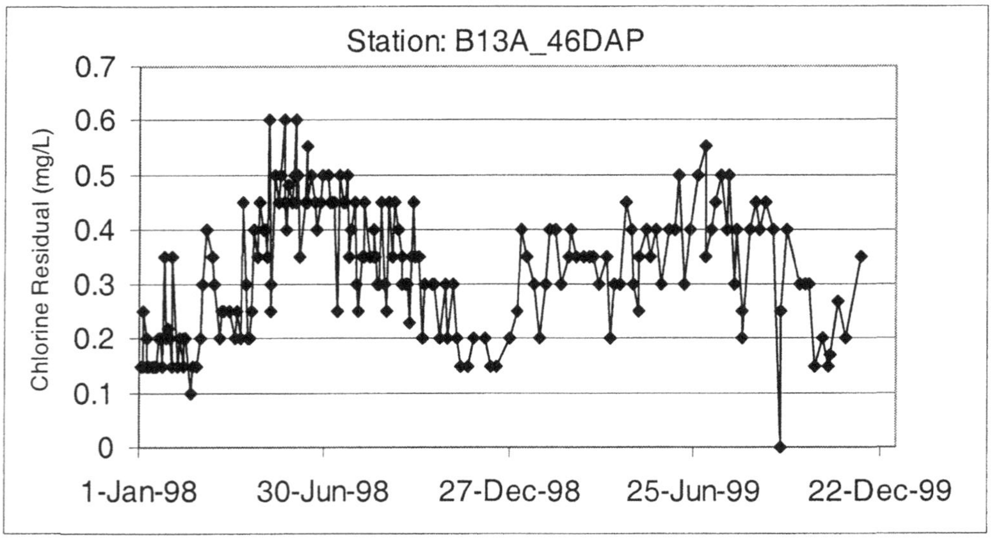
Station: B13A_46DAP
Chlorine Residual (mg/L)
0.7
0.6
0.5
0.4
0.3
0.2
0.1
0
1-Jan-98
30-Jun-98
27-Dec-98
25-Jun-99
22-Dec-99

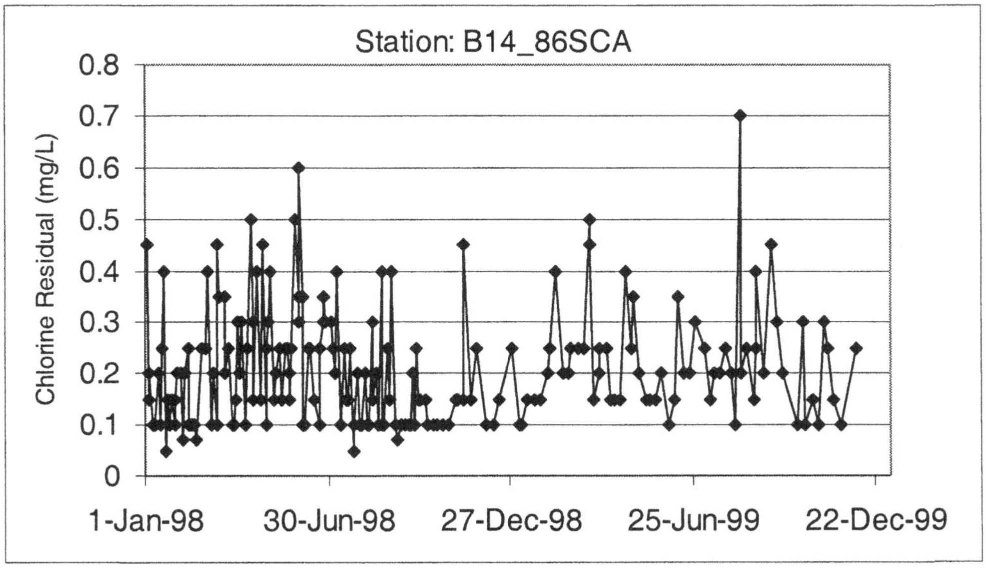
Station: B14_86SCA
Chlorine Residual (mg/L)
0.8
0.7
0.6
0.5
0.4
0.3
0.2
0.1
0
1-Jan-98
30-Jun-98
27-Dec-98
25-Jun-99
22-Dec-99

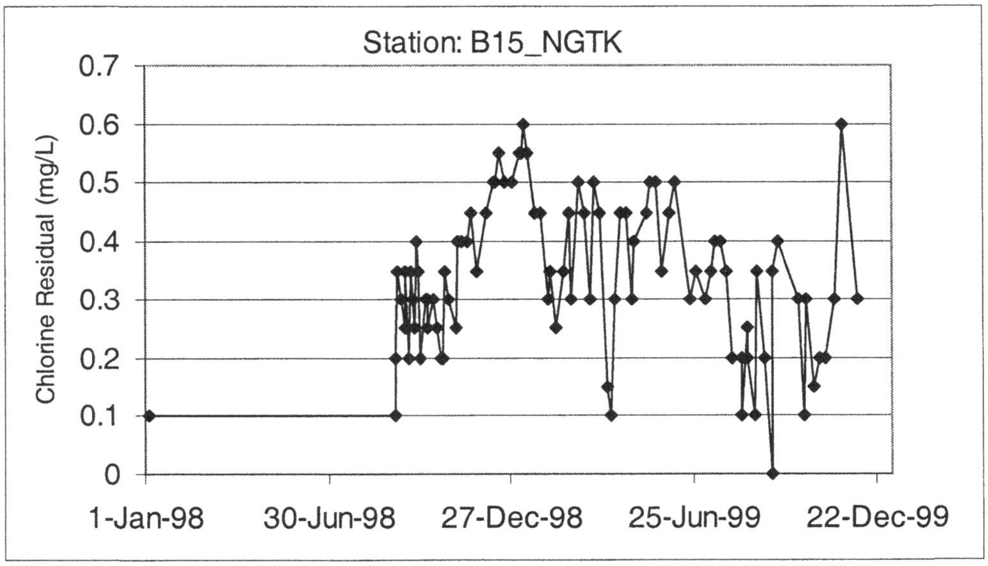
Station: B15_NGTK
Chlorine Residual (mg/L)
0.7
0.6
0.5
0.4
0.3
0.2
0.1
0
1-Jan-98
30-Jun-98
27-Dec-98
25-Jun-99
22-Dec-99

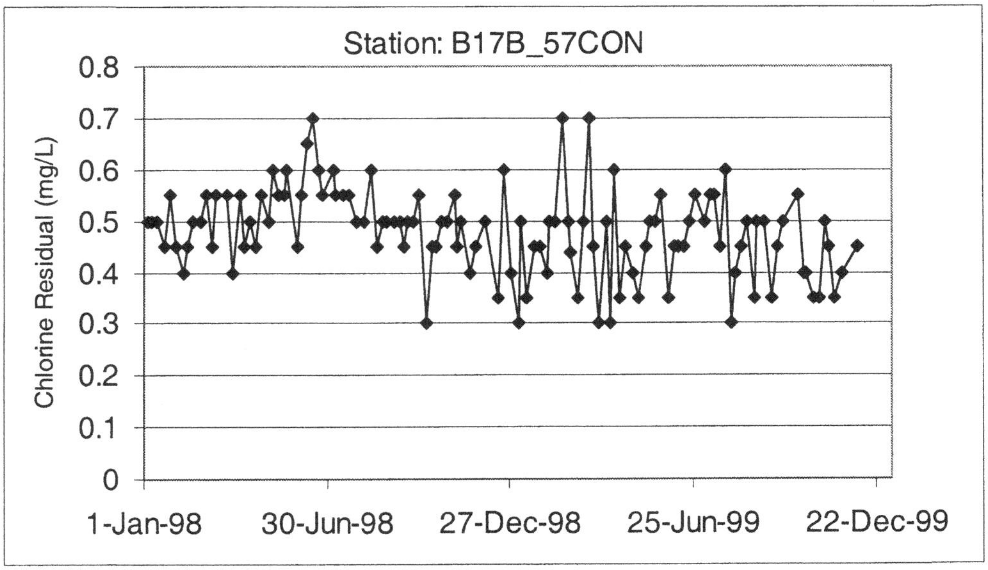
Station: B17B_57CON
Chlorine Residual (mg/L)
0.8
0.7
0.6
0.5
0.4
0.3
0.2
0.1
0
1-Jan-98
30-Jun-98
27-Dec-98
25-Jun-99
22-Dec-99

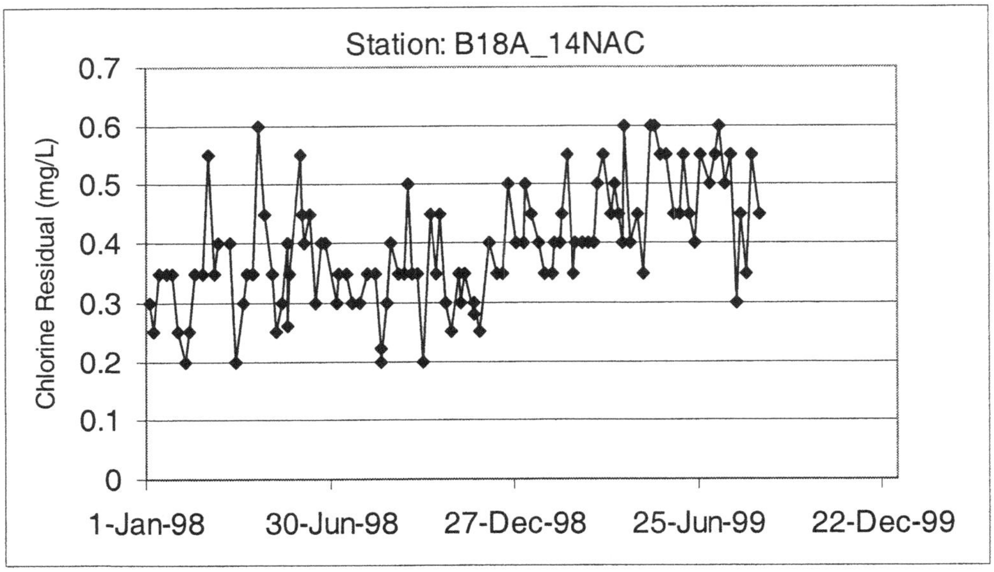
Station: B18A_14NAC
Chlorine Residual (mg/L)
0.7
0.6
0.5
0.4
0.3
0.2
0.1
0
1-Jan-98
30-Jun-98
27-Dec-98
25-Jun-99
22-Dec-99

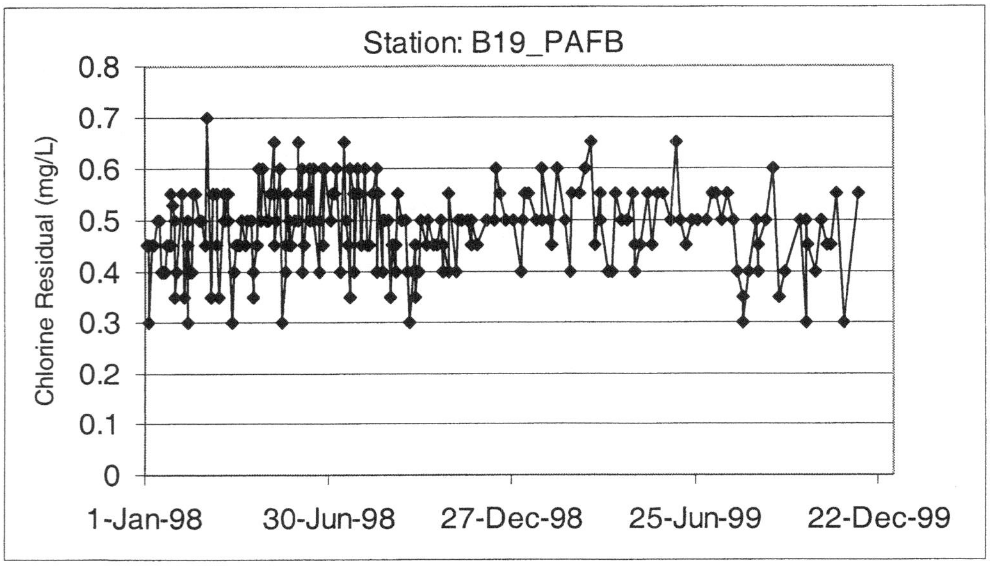
Station: B19_PAFB
Chlorine Residual (mg/L)
0.8
0.7
0.6
0.5
0.4
0.3
0.2
0.1
0
1-Jan-98
30-Jun-98
27-Dec-98
25-Jun-99
22-Dec-99

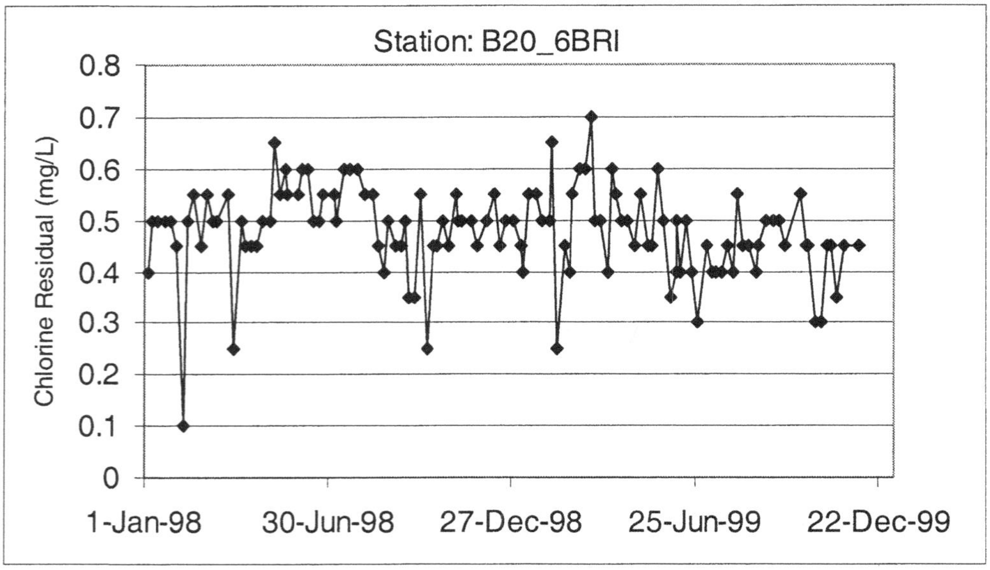
Station: B20_6BRI
Chlorine Residual (mg/L)
0.8
0.7
0.6
0.5
0.4
0.3
0.2
0.1
0
1-Jan-98
30-Jun-98
27-Dec-98
25-Jun-99
22-Dec-99

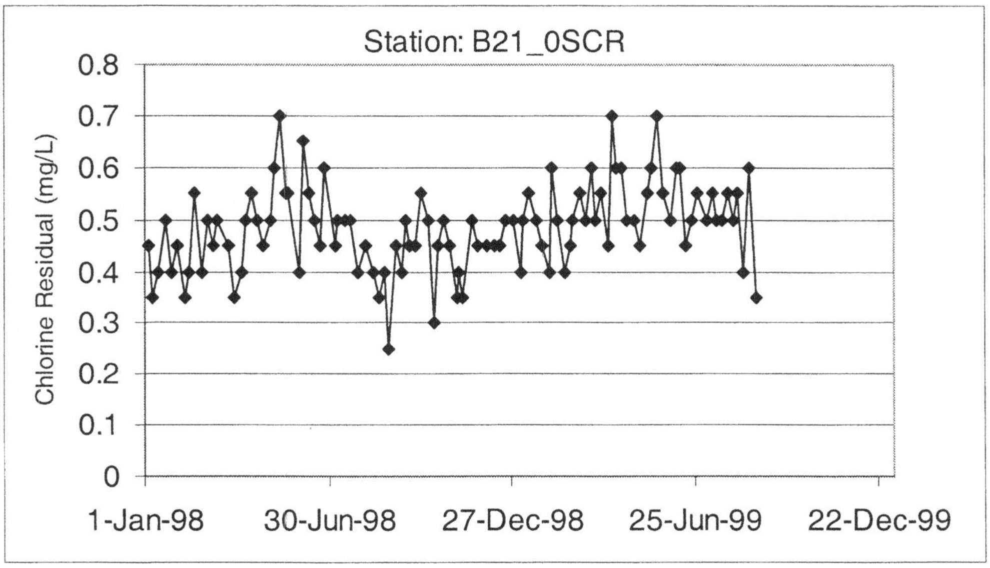
Station: B21_0SCR
Chlorine Residual (mg/L)
0.8
0.7
0.6
0.5
0.4
0.3
0.2
0.1
0
1-Jan-98
30-Jun-98
27-Dec-98
25-Jun-99
22-Dec-99

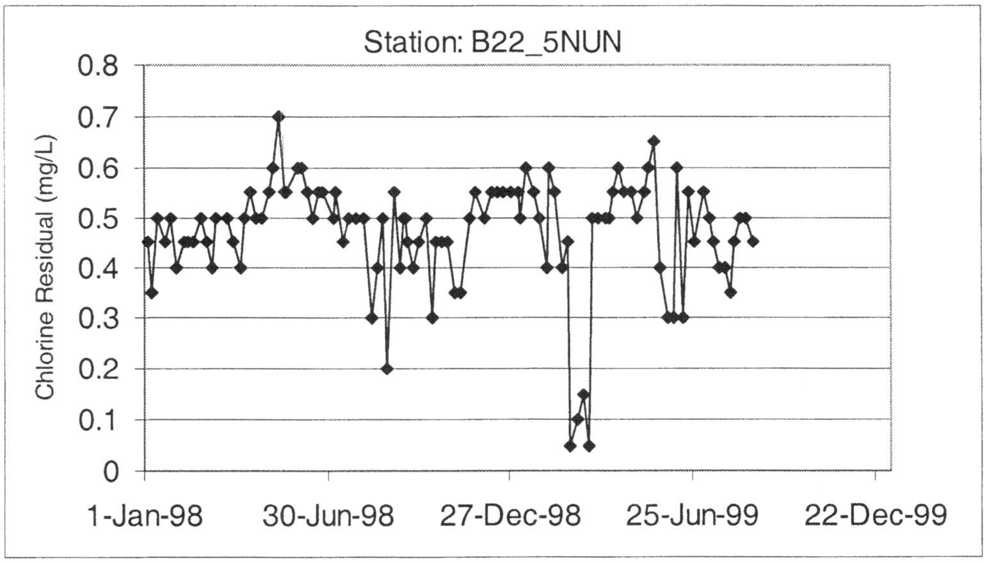
Station: B22_5NUN
Chlorine Residual (mg/L)
0.8
0.7
0.6
0.5
0.4
0.3
0.2
0.1
0
1-Jan-98
30-Jun-98
27-Dec-98
25-Jun-99
22-Dec-99

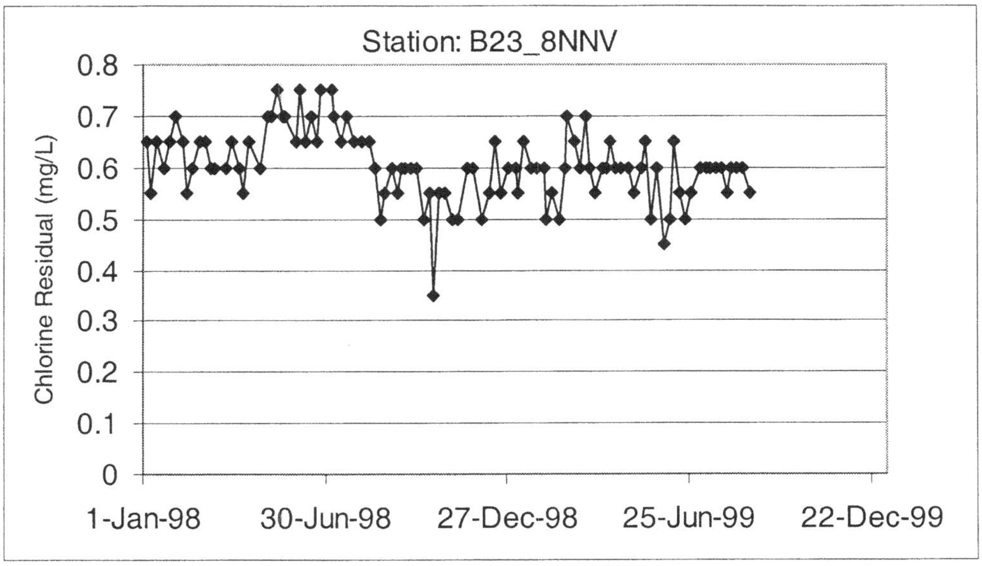
Station: B23_8NNV
Chlorine Residual (mg/L)
0.8
0.7
0.6
0.5
0.4
0.3
0.2
0.1
0
1-Jan-98
30-Jun-98
27-Dec-98
25-Jun-99
22-Dec-99

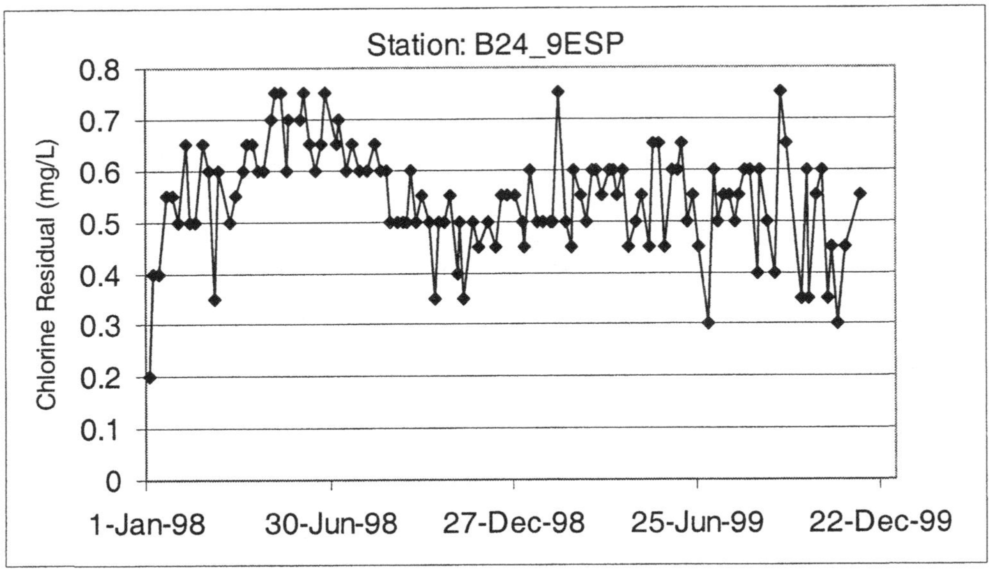
Station: B24_9ESP
Chlorine Residual (mg/L)
0.8
0.7
0.6
0.5
0.4
0.3
0.2
0.1
0
1-Jan-98
30-Jun-98
27-Dec-98
25-Jun-99
22-Dec-99

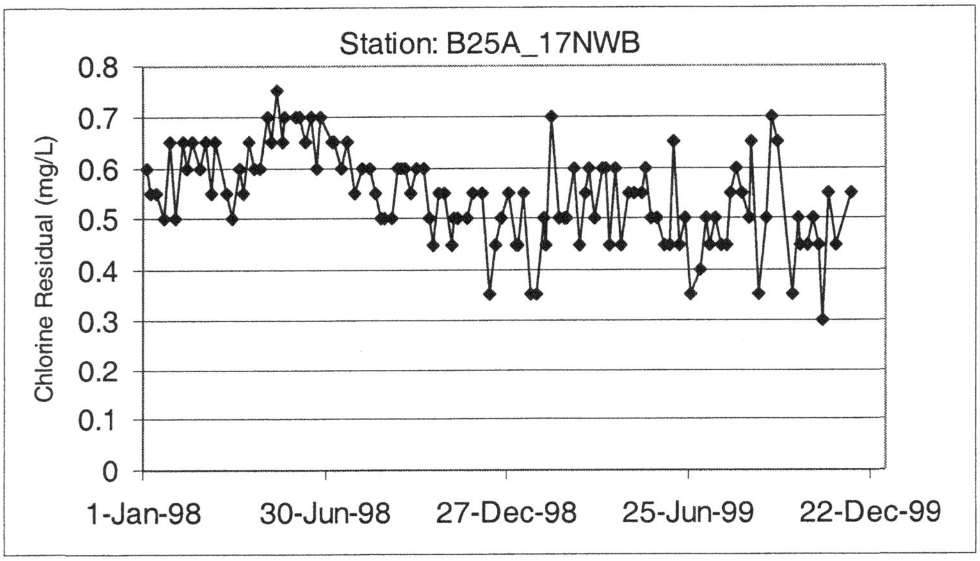
Station: B25A_17NWB
Chlorine Residual (mg/L)
0.8
0.7
0.6
0.5
0.4
0.3
0.2
0.1
0
1-Jan-98
30-Jun-98
27-Dec-98
25-Jun-99
22-Dec-99

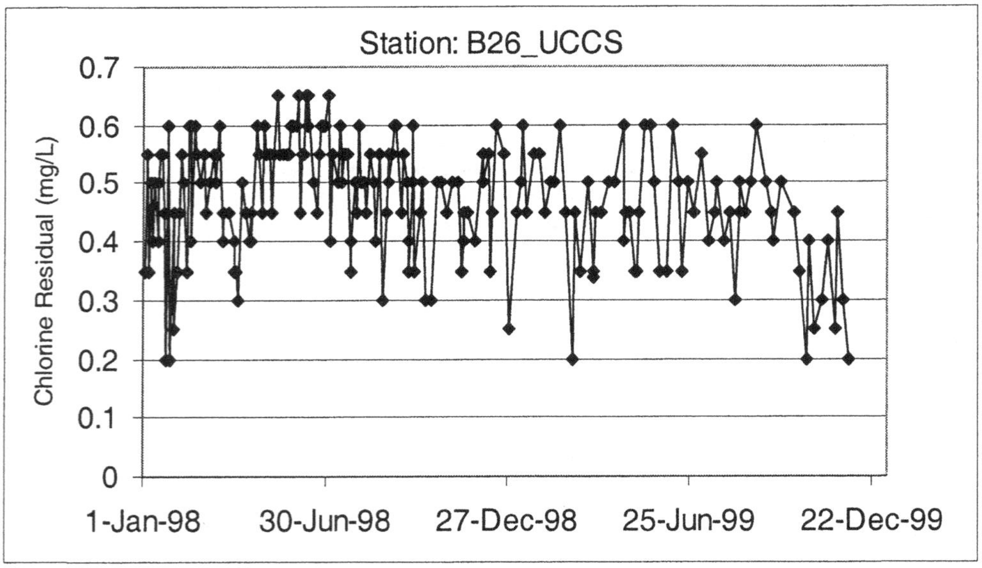
Station: B26_UCCS
Chlorine Residual (mg/L)
0.7
0.6
0.5
0.4
0.3
0.2
0.1
0
1-Jan-98
30-Jun-98
27-Dec-98
25-Jun-99
22-Dec-99

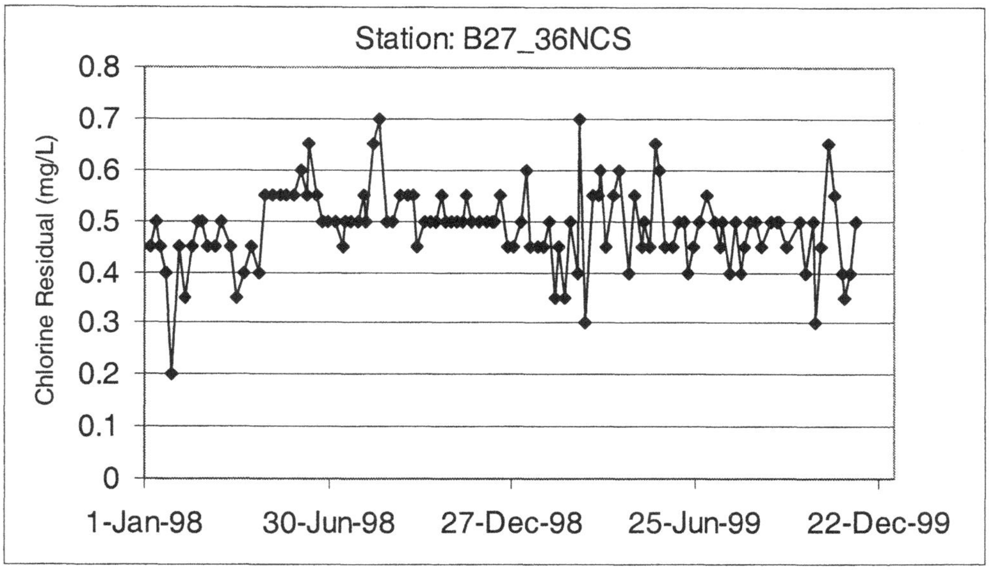
Station: B27_36NCS
Chlorine Residual (mg/L)
0.8
0.7
0.6
0.5
0.4
0.3
0.2
0.1
0
1-Jan-98
30-Jun-98
27-Dec-98
25-Jun-99
22-Dec-99

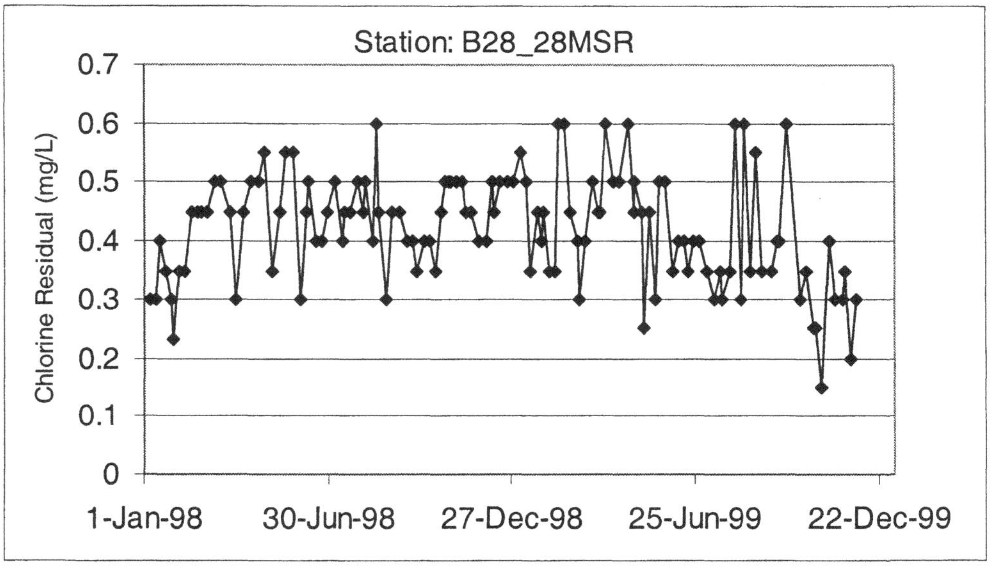
Station: B28_28MSR
Chlorine Residual (mg/L)
0.7
0.6
0.5
0.4
0.3
0.2
0.1
0
1-Jan-98
30-Jun-98
27-Dec-98
25-Jun-99
22-Dec-99

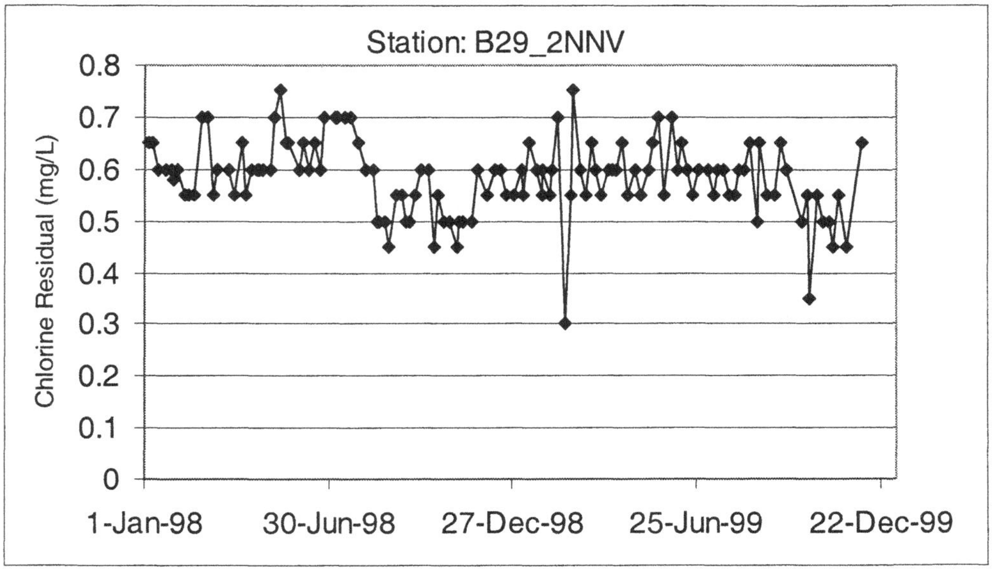
Station: B29_2NNV
Chlorine Residual (mg/L)
0.8
0.7
0.6
0.5
0.4
0.3
0.2
0.1
0
1-Jan-98
30-Jun-98
27-Dec-98
25-Jun-99
22-Dec-99

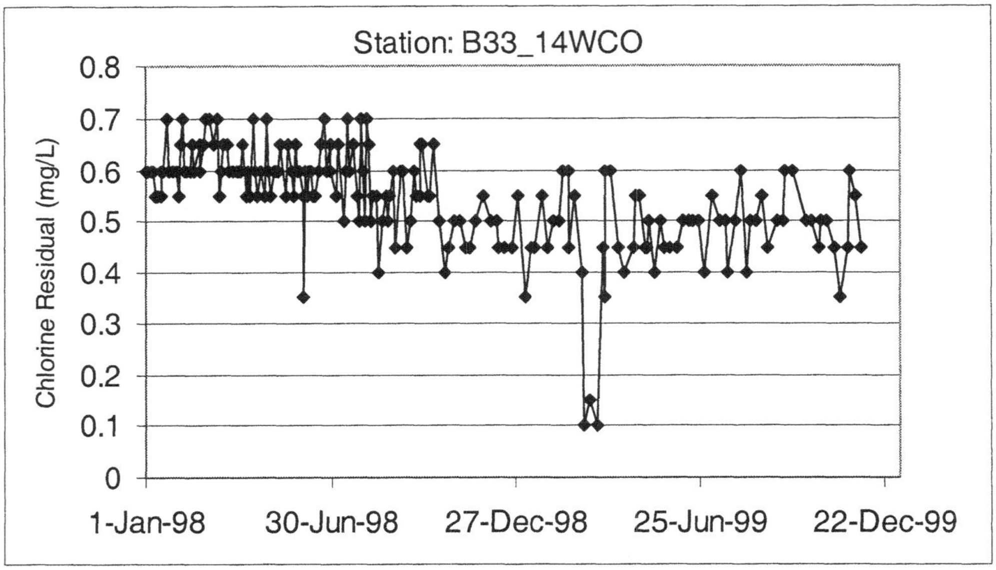
Station: B33_14WCO
Chlorine Residual (mg/L)
0.8
0.7
0.6
0.5
0.4
0.3
0.2
0.1
0
1-Jan-98
30-Jun-98
27-Dec-98
25-Jun-99
22-Dec-99

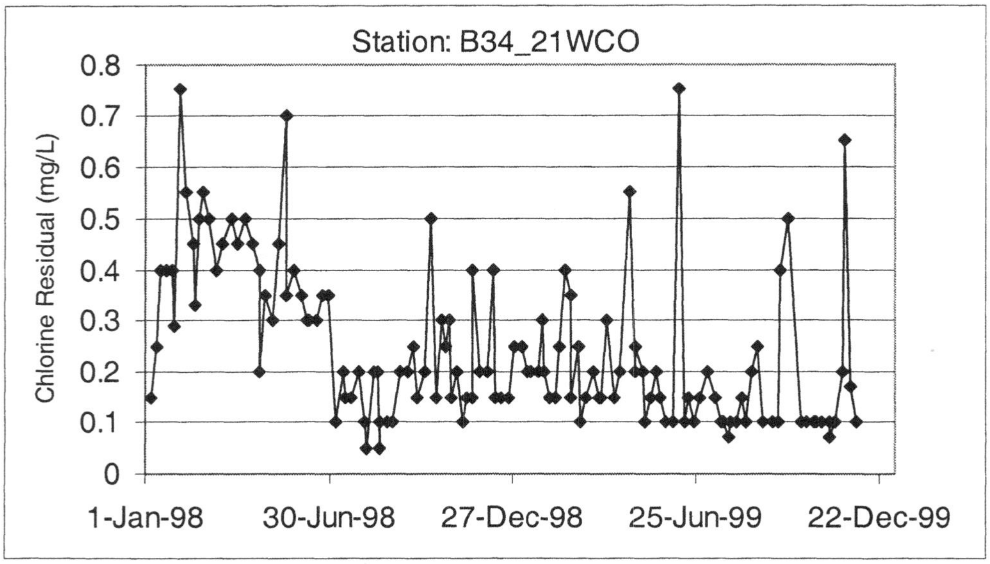
Station: B34_21WCO
Chlorine Residual (mg/L)
0.8
0.7
0.6
0.5
0.4
0.3
0.2
0.1
0
1-Jan-98
30-Jun-98
27-Dec-98
25-Jun-99
22-Dec-99

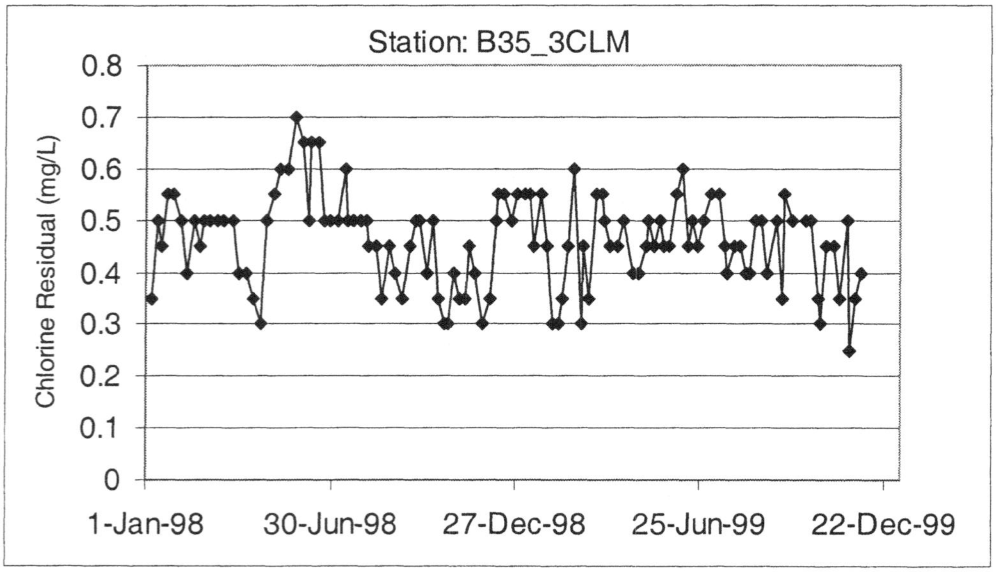
Station: B35_3CLM
Chlorine Residual (mg/L)
0.8
0.7
0.6
0.5
0.4
0.3
0.2
0.1
0
1-Jan-98
30-Jun-98
27-Dec-98
25-Jun-99
22-Dec-99

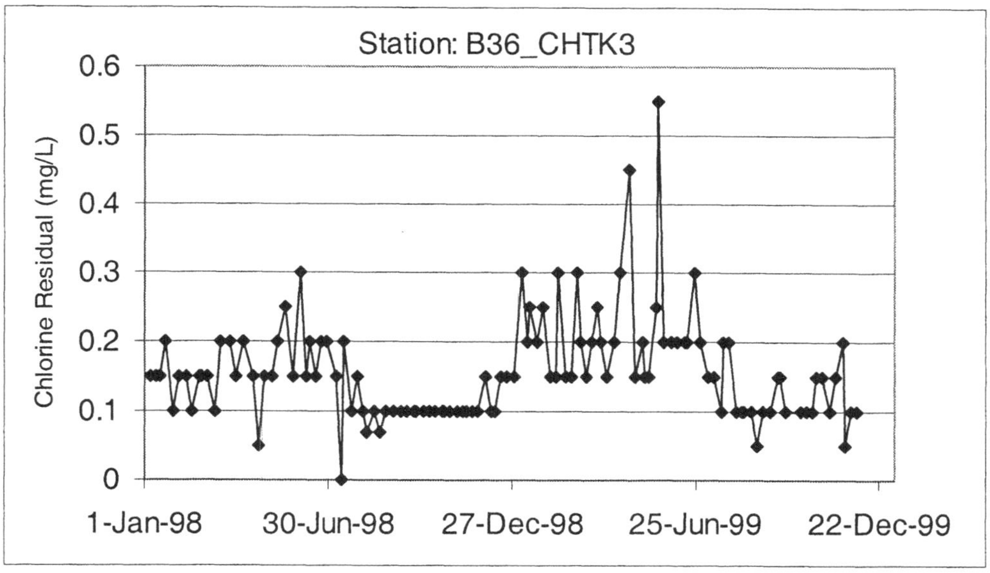
Station: B36_CHTK3
Chlorine Residual (mg/L)
0
0.1
0.2
0.3
0.4
0.5
0.6
1-Jan-98
30-Jun-98
27-Dec-98
25-Jun-99
22-Dec-99

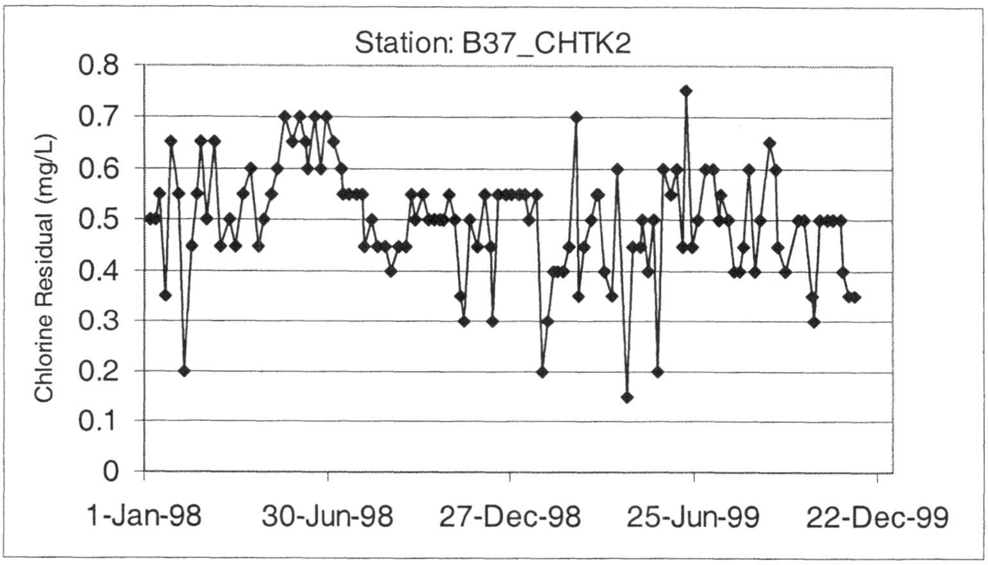
Station: B37_CHTK2
Chlorine Residual (mg/L)
0
0.1
0.2
0.3
0.4
0.5
0.6
0.7
0.8
1-Jan-98
30-Jun-98
27-Dec-98
25-Jun-99
22-Dec-99

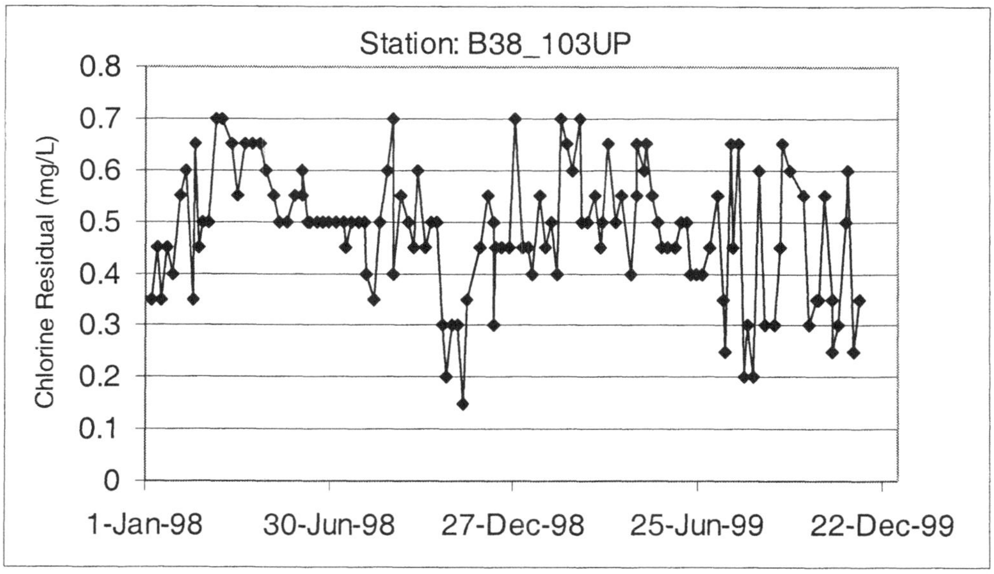
Station: B38_103UP
Chlorine Residual (mg/L)
0
0.1
0.2
0.3
0.4
0.5
0.6
0.7
0.8
1-Jan-98
30-Jun-98
27-Dec-98
25-Jun-99
22-Dec-99

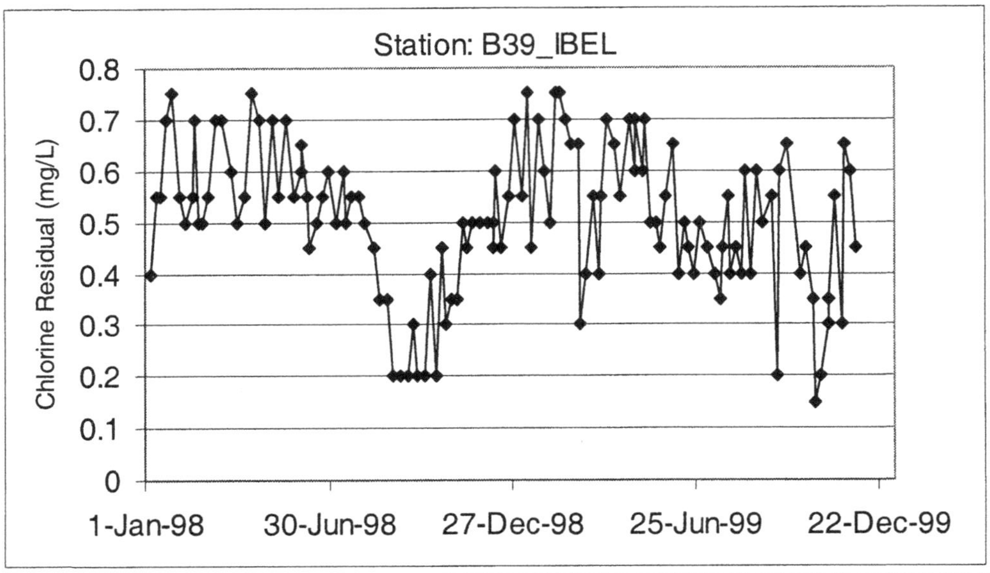
Station: B39_IBEL
Chlorine Residual (mg/L)
0.8
0.7
0.6
0.5
0.4
0.3
0.2
0.1
0
1-Jan-98
30-Jun-98
27-Dec-98
25-Jun-99
22-Dec-99

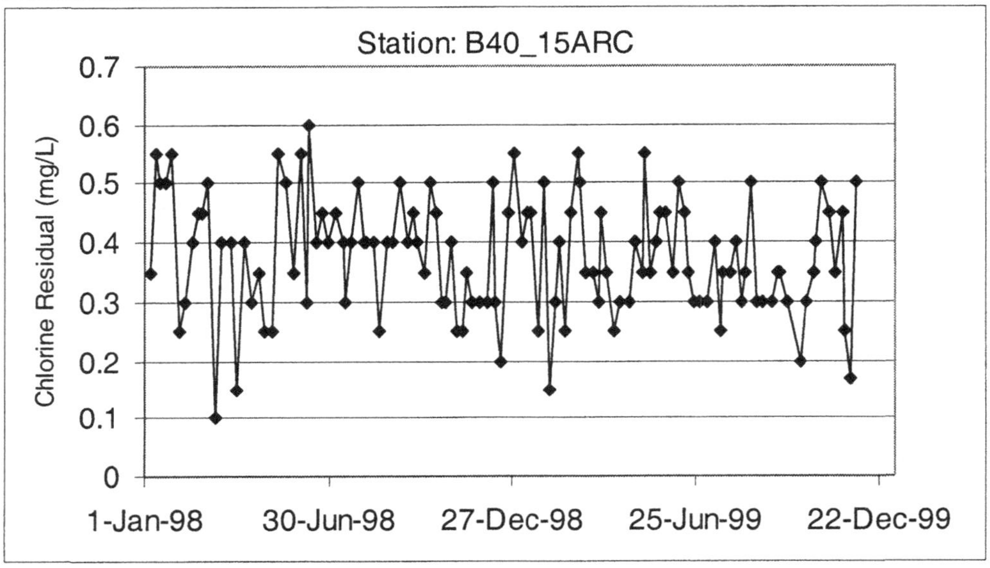
Station: B40_15ARC
Chlorine Residual (mg/L)
0.7
0.6
0.5
0.4
0.3
0.2
0.1
0
1-Jan-98
30-Jun-98
27-Dec-98
25-Jun-99
22-Dec-99

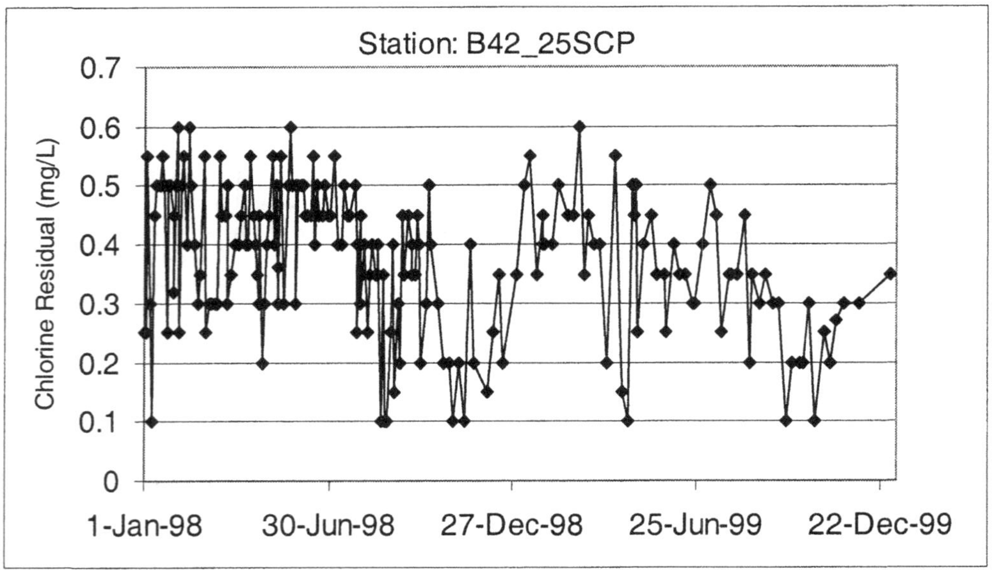
Station: B42_25SCP
Chlorine Residual (mg/L)
0.7
0.6
0.5
0.4
0.3
0.2
0.1
0
1-Jan-98
30-Jun-98
27-Dec-98
25-Jun-99
22-Dec-99

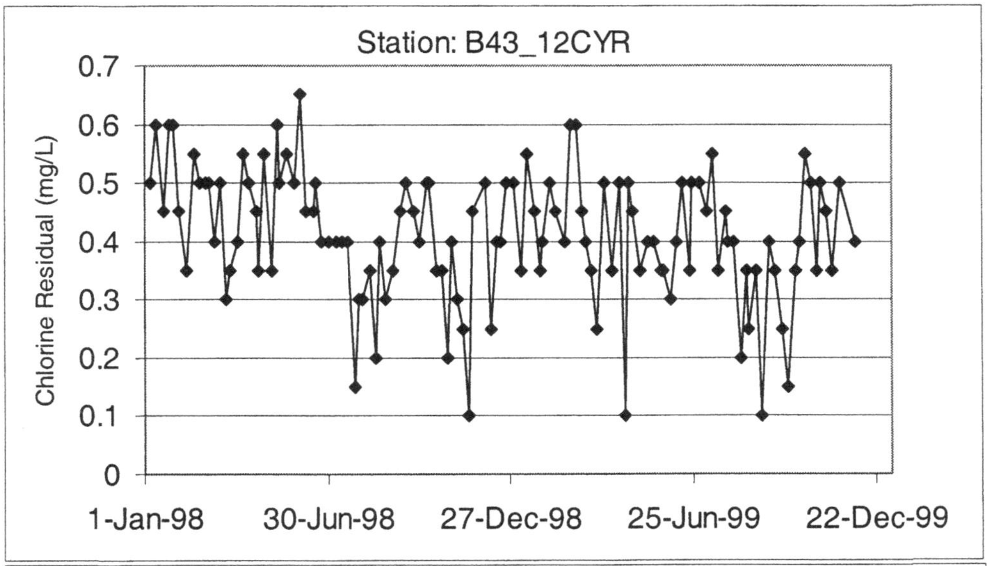
Station: B43_12CYR
Chlorine Residual (mg/L)
0.7
0.6
0.5
0.4
0.3
0.2
0.1
0
1-Jan-98
30-Jun-98
27-Dec-98
25-Jun-99
22-Dec-99

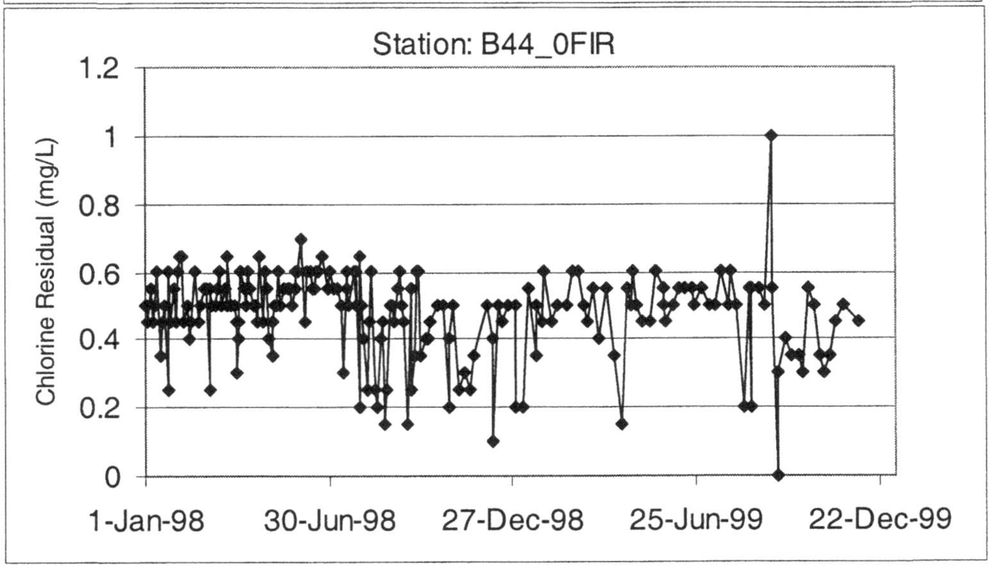
Station: B44_0FIR
Chlorine Residual (mg/L)
1.2
1
0.8
0.6
0.4
0.2
0
1-Jan-98
30-Jun-98
27-Dec-98
25-Jun-99
22-Dec-99

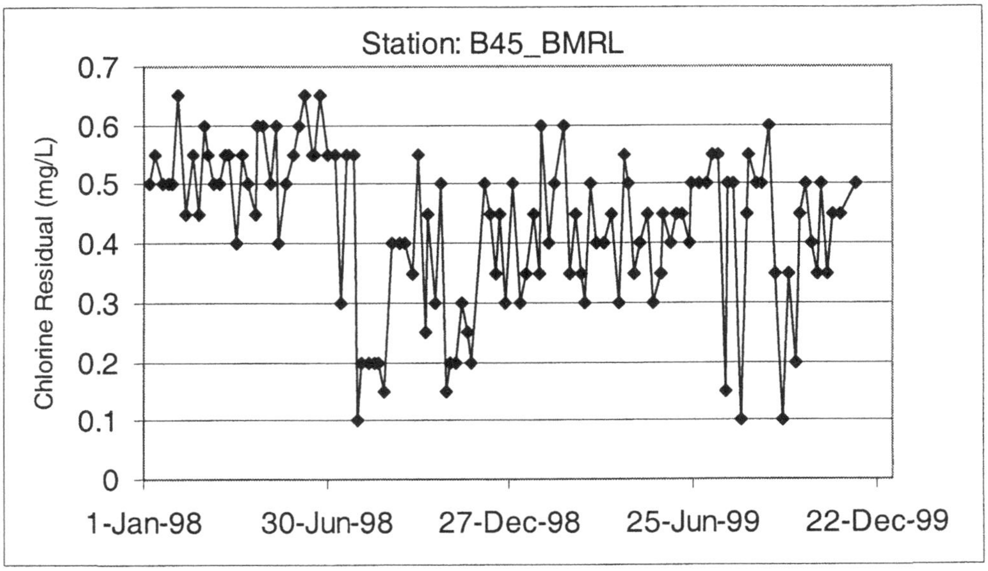
Station: B45_BMRL
Chlorine Residual (mg/L)
0.7
0.6
0.5
0.4
0.3
0.2
0.1
0
1-Jan-98
30-Jun-98
27-Dec-98
25-Jun-99
22-Dec-99

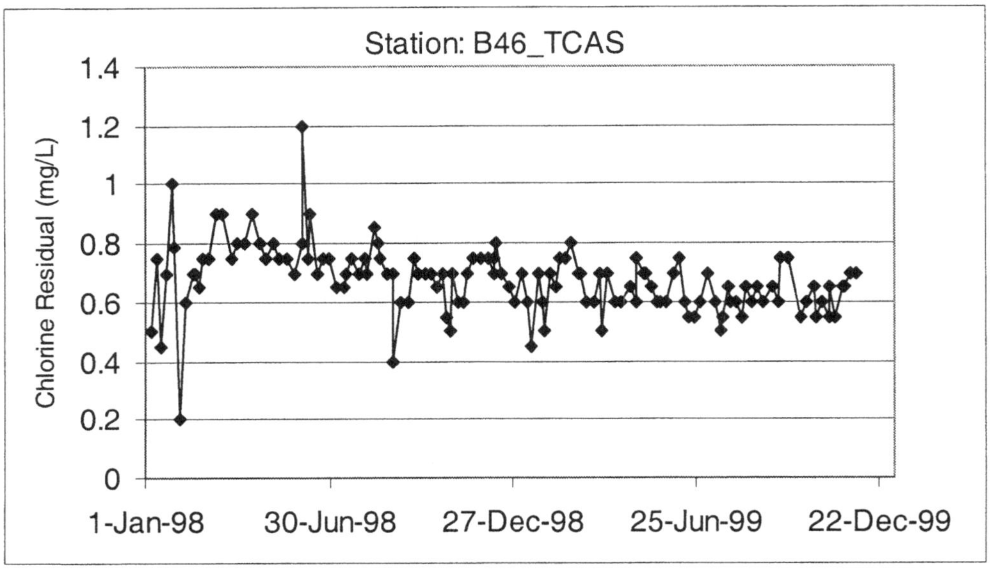
Station: B46_TCAS
Chlorine Residual (mg/L)
1.4
1.2
1
0.8
0.6
0.4
0.2
0
1-Jan-98
30-Jun-98
27-Dec-98
25-Jun-99
22-Dec-99

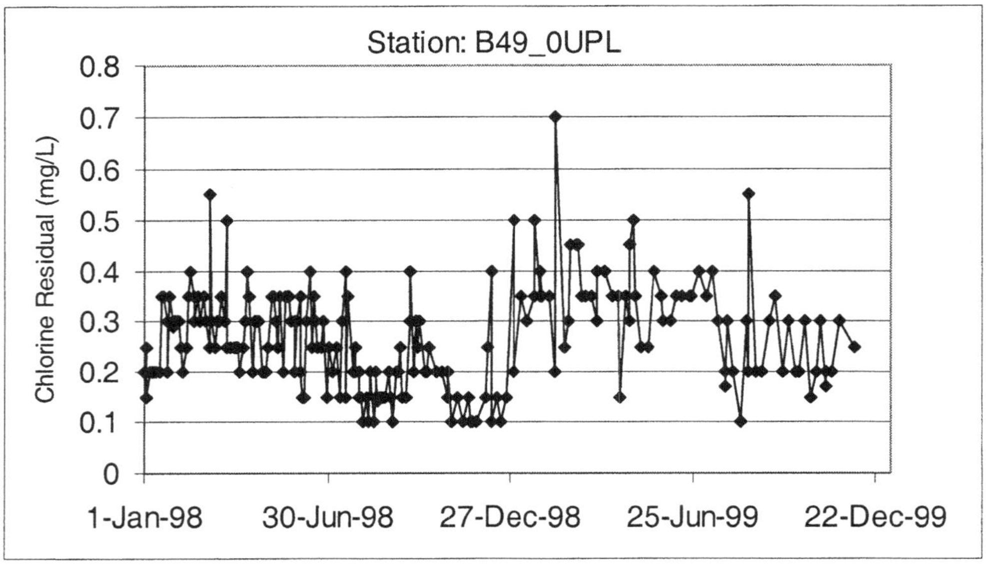
Station: B49_0UPL
Chlorine Residual (mg/L)
0.8
0.7
0.6
0.5
0.4
0.3
0.2
0.1
0
1-Jan-98
30-Jun-98
27-Dec-98
25-Jun-99
22-Dec-99

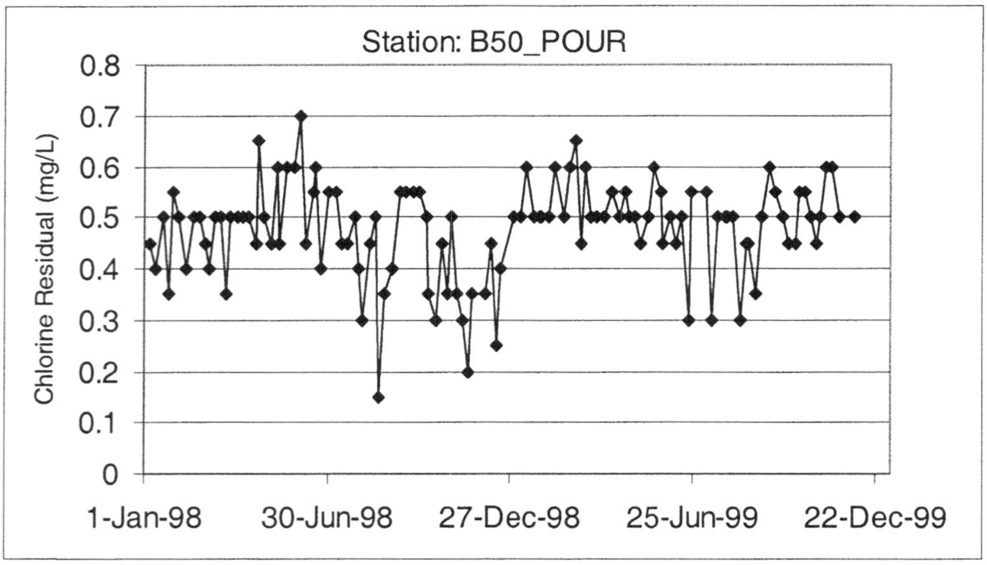
Station: B50_POUR
Chlorine Residual (mg/L)
0.8
0.7
0.6
0.5
0.4
0.3
0.2
0.1
0
1-Jan-98
30-Jun-98
27-Dec-98
25-Jun-99
22-Dec-99

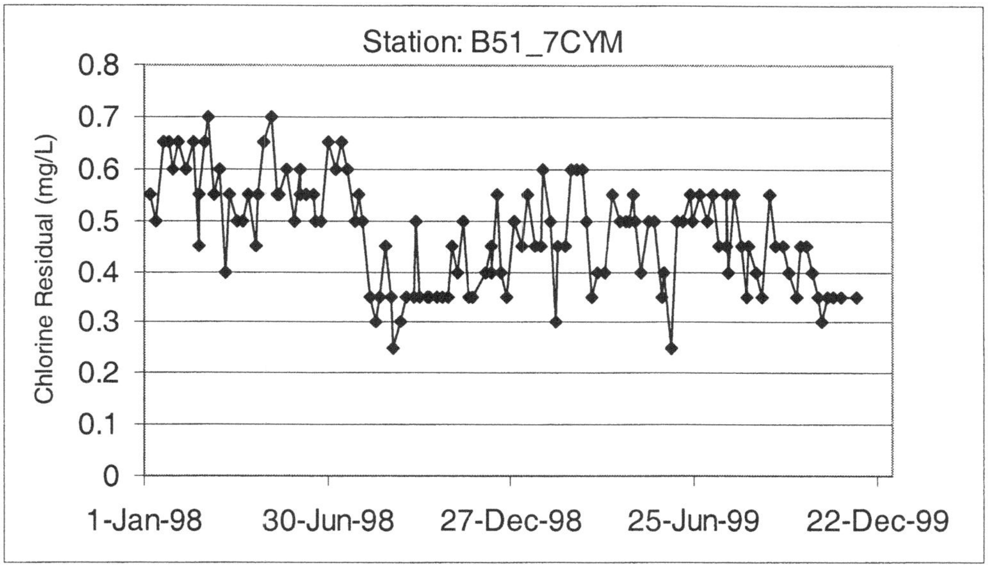
Station: B51_7CYM
Chlorine Residual (mg/L)
0.8
0.7
0.6
0.5
0.4
0.3
0.2
0.1
0
1-Jan-98
30-Jun-98
27-Dec-98
25-Jun-99
22-Dec-99

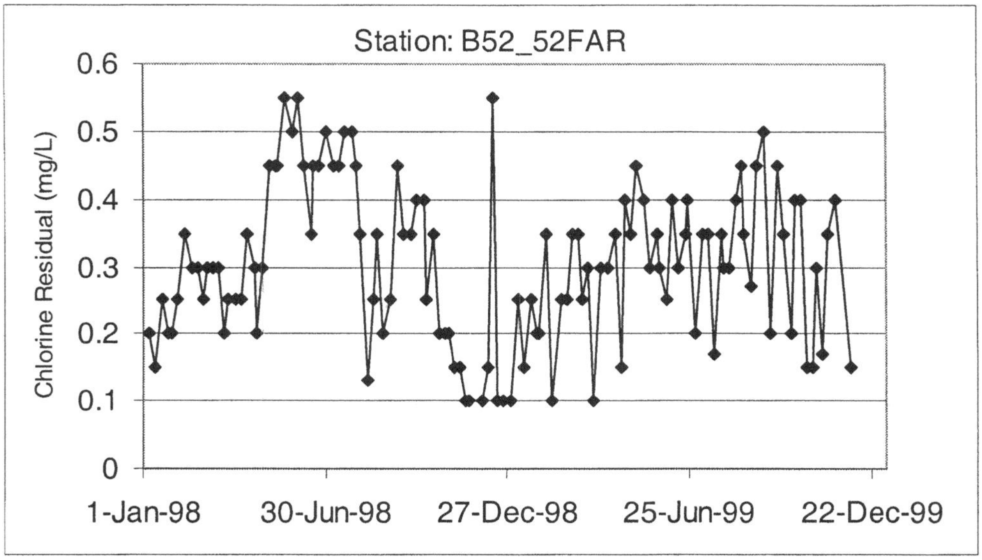
Station: B52_52FAR
Chlorine Residual (mg/L)
0.6
0.5
0.4
0.3
0.2
0.1
0
1-Jan-98
30-Jun-98
27-Dec-98
25-Jun-99
22-Dec-99

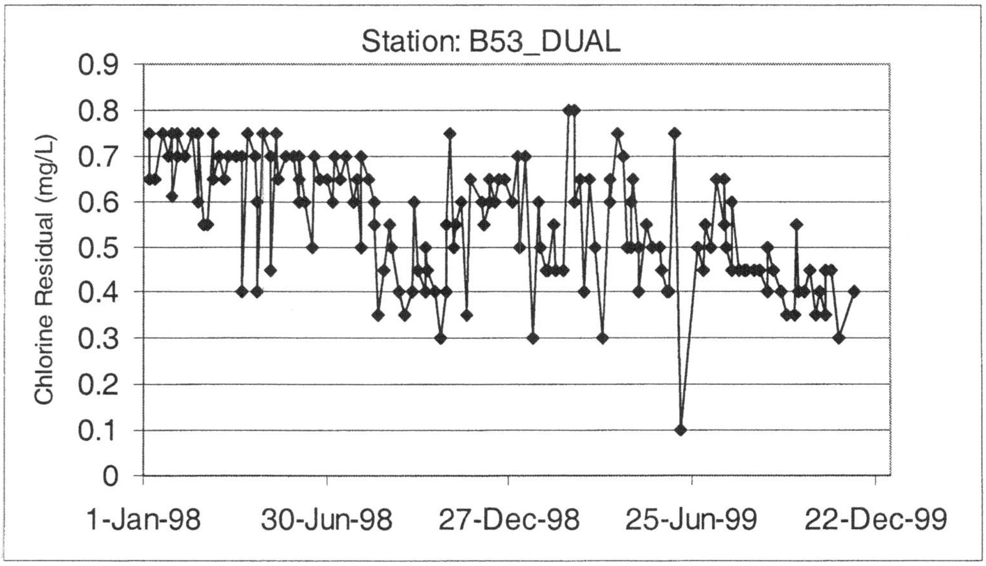
Station: B53_DUAL
Chlorine Residual (mg/L)
0.9
0.8
0.7
0.6
0.5
0.4
0.3
0.2
0.1
0
1-Jan-98
30-Jun-98
27-Dec-98
25-Jun-99
22-Dec-99

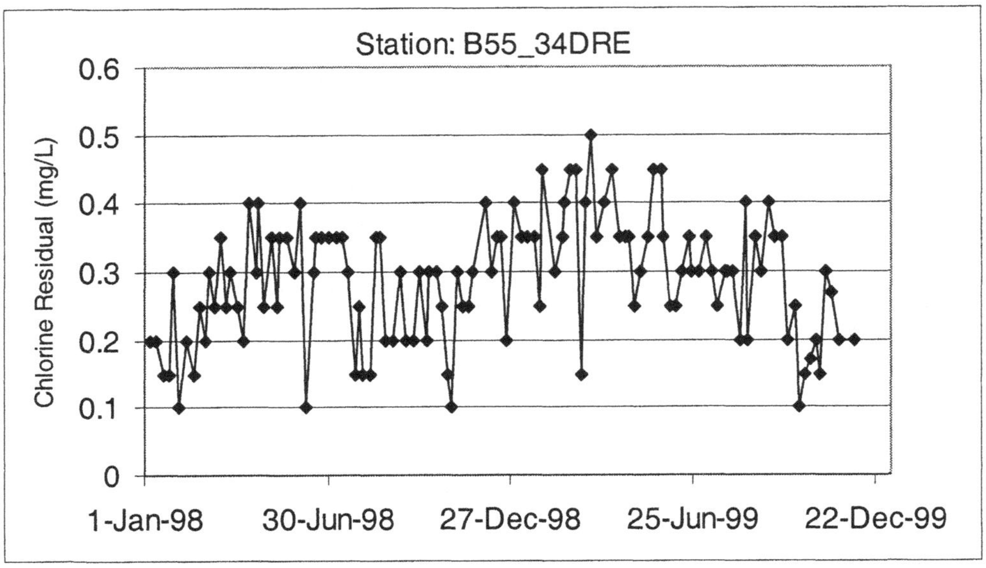
Station: B55_34DRE
Chlorine Residual (mg/L)
0.6
0.5
0.4
0.3
0.2
0.1
0
1-Jan-98
30-Jun-98
27-Dec-98
25-Jun-99
22-Dec-99

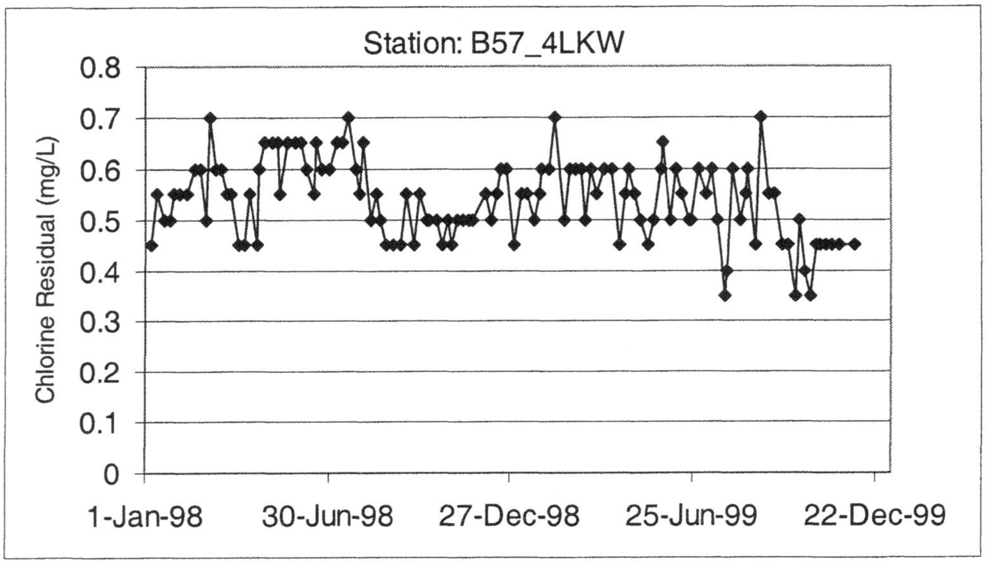
Station: B57_4LKW
Chlorine Residual (mg/L)
0.8
0.7
0.6
0.5
0.4
0.3
0.2
0.1
0
1-Jan-98
30-Jun-98
27-Dec-98
25-Jun-99
22-Dec-99

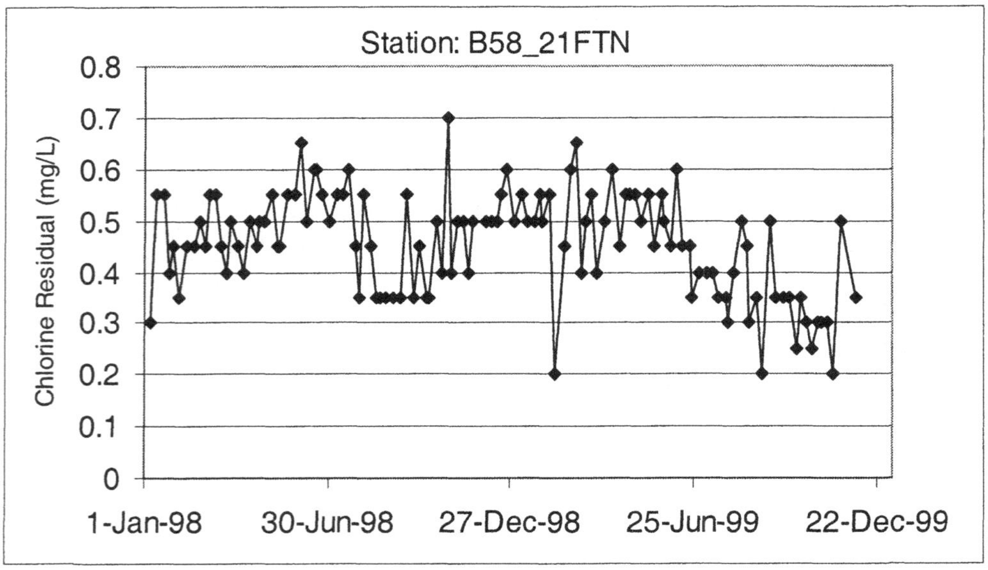
Station: B58_21FTN
Chlorine Residual (mg/L)
0.8
0.7
0.6
0.5
0.4
0.3
0.2
0.1
0
1-Jan-98
30-Jun-98
27-Dec-98
25-Jun-99
22-Dec-99

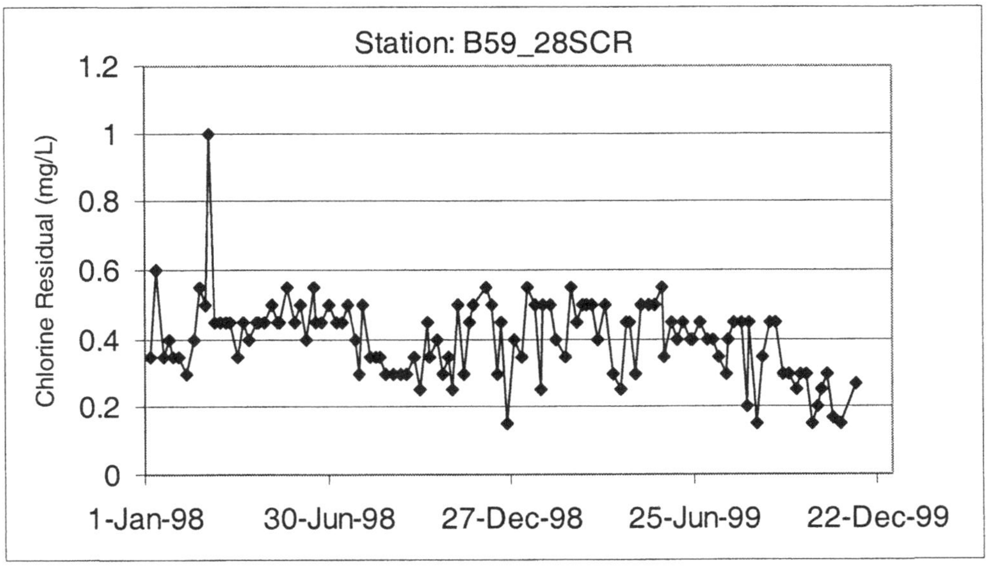
Station: B59_28SCR
Chlorine Residual (mg/L)
1.2
1
0.8
0.6
0.4
0.2
0
1-Jan-98
30-Jun-98
27-Dec-98
25-Jun-99
22-Dec-99

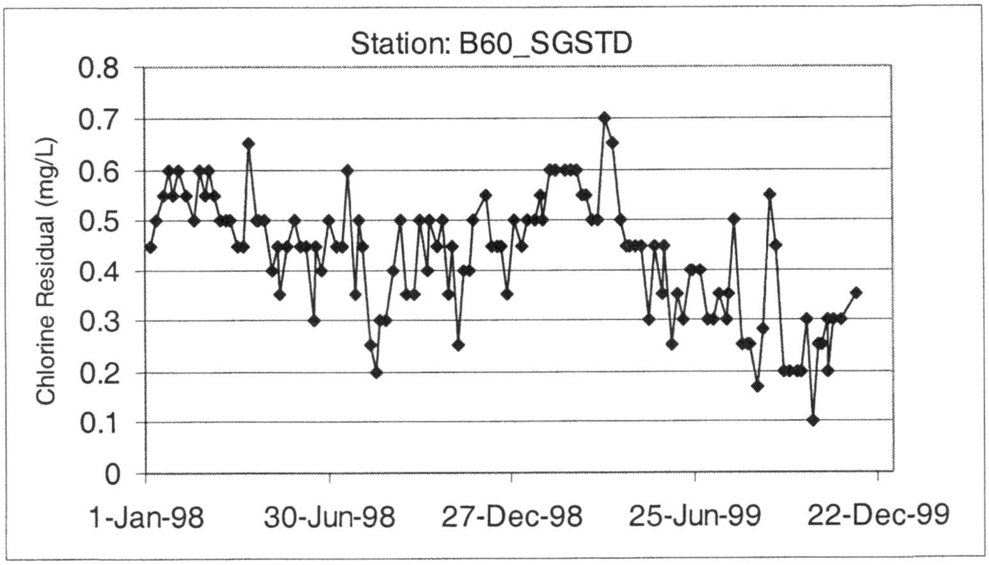
Station: B60_SGSTD
Chlorine Residual (mg/L)
0.8
0.7
0.6
0.5
0.4
0.3
0.2
0.1
0
1-Jan-98
30-Jun-98
27-Dec-98
25-Jun-99
22-Dec-99

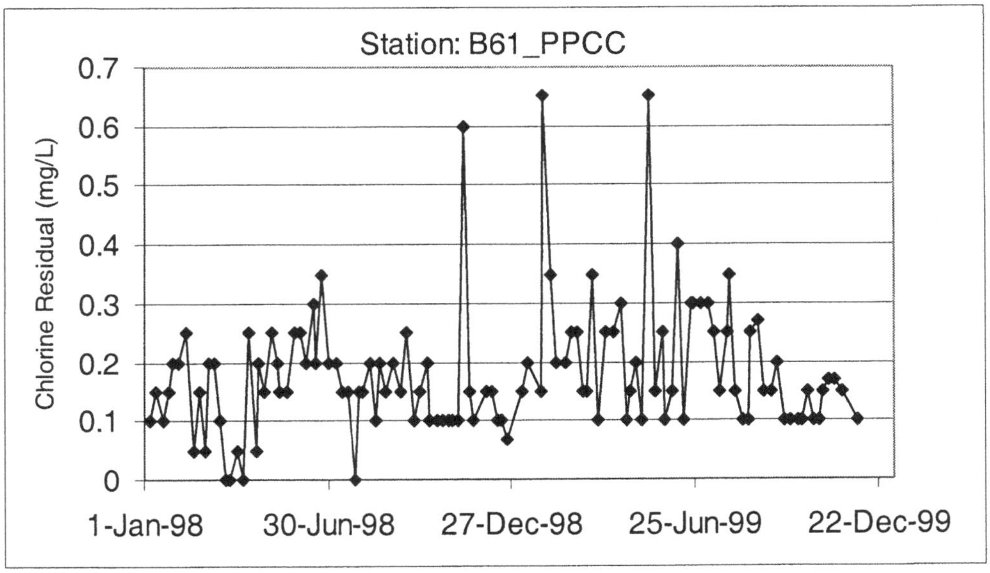
Station: B61_PPCC
Chlorine Residual (mg/L)
0.7
0.6
0.5
0.4
0.3
0.2
0.1
0
1-Jan-98
30-Jun-98
27-Dec-98
25-Jun-99
22-Dec-99

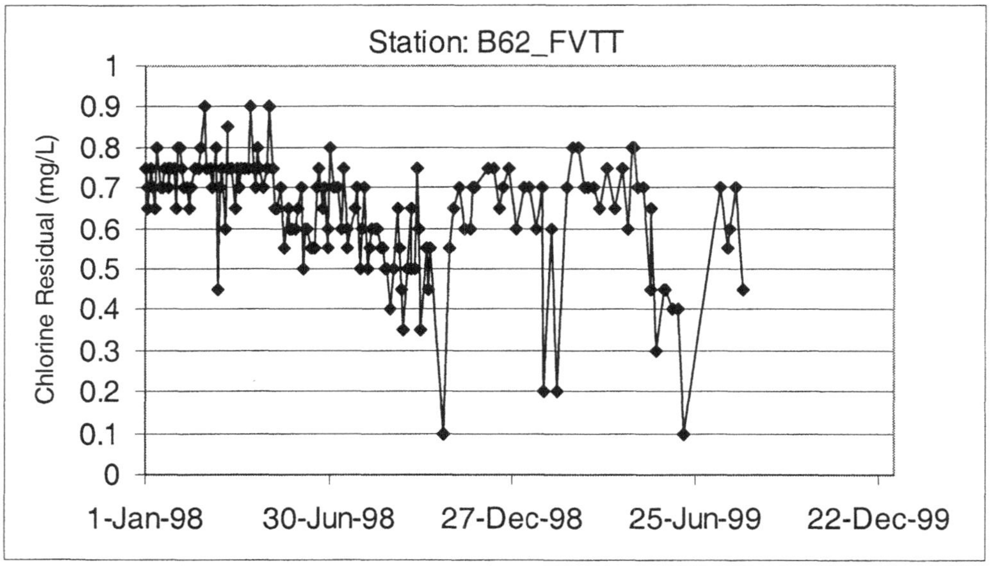
Station: B62_FVTT
Chlorine Residual (mg/L)
1
0.9
0.8
0.7
0.6
0.5
0.4
0.3
0.2
0.1
0
1-Jan-98
30-Jun-98
27-Dec-98
25-Jun-99
22-Dec-99

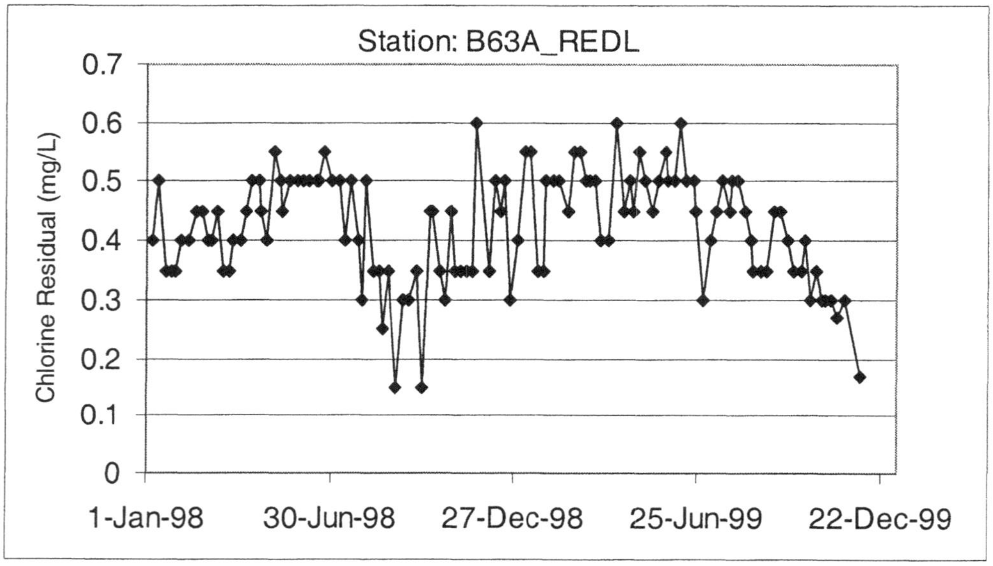
Station: B63A_REDL
Chlorine Residual (mg/L)
0.7
0.6
0.5
0.4
0.3
0.2
0.1
0
1-Jan-98
30-Jun-98
27-Dec-98
25-Jun-99
22-Dec-99

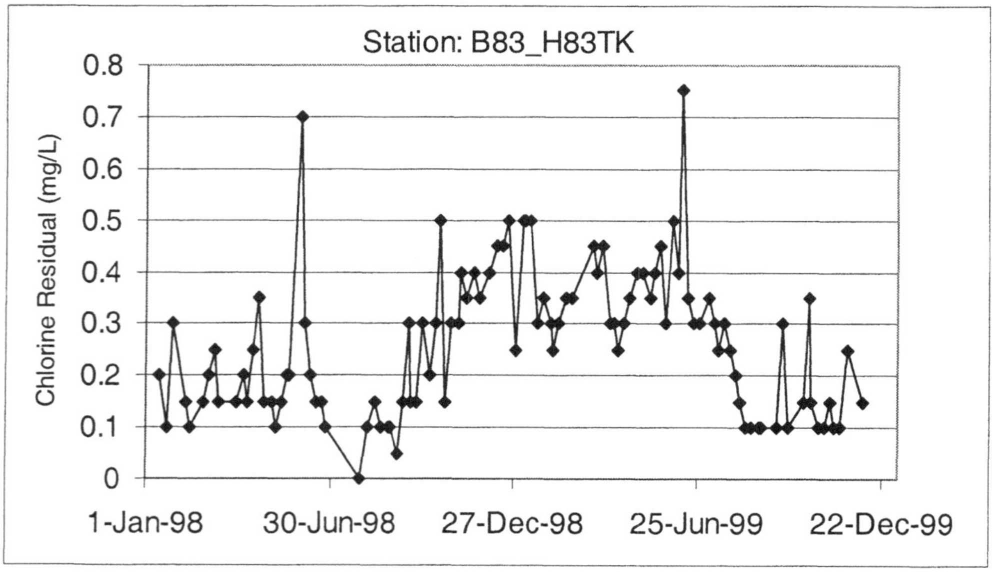
Station: B83_H83TK
Chlorine Residual (mg/L)
0.8
0.7
0.6
0.5
0.4
0.3
0.2
0.1
0
1-Jan-98
30-Jun-98
27-Dec-98
25-Jun-99
22-Dec-99

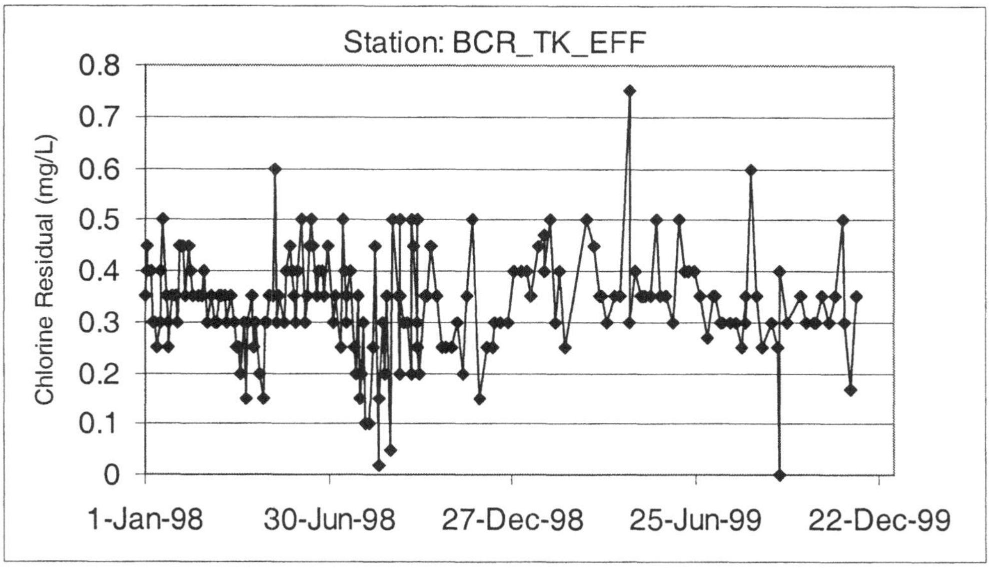
Station: BCR_TK_EFF
Chlorine Residual (mg/L)
0.8
0.7
0.6
0.5
0.4
0.3
0.2
0.1
0
1-Jan-98
30-Jun-98
27-Dec-98
25-Jun-99
22-Dec-99

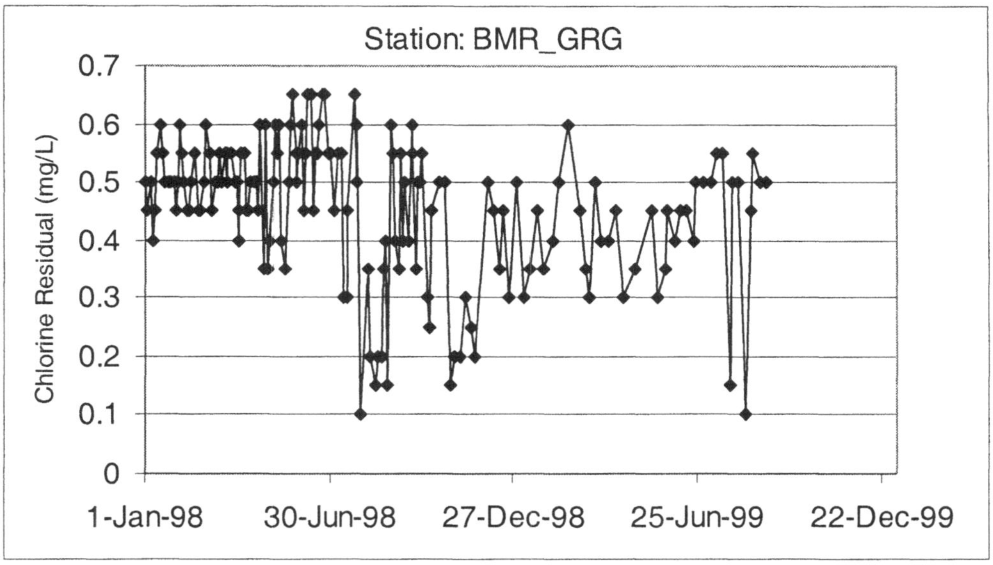
Station: BMR_GRG
Chlorine Residual (mg/L)
0.7
0.6
0.5
0.4
0.3
0.2
0.1
0
1-Jan-98
30-Jun-98
27-Dec-98
25-Jun-99
22-Dec-99

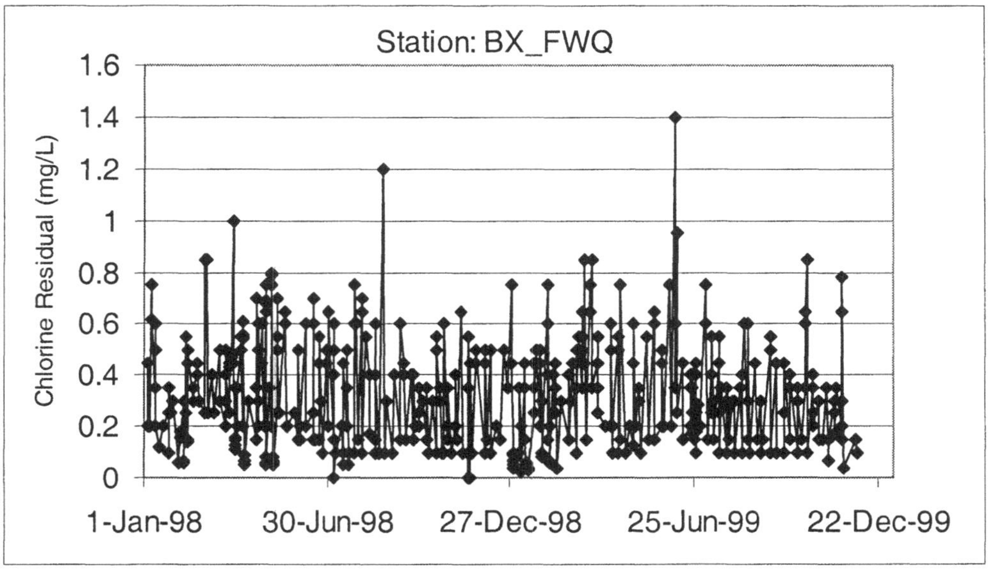
Station: BX_FWQ
Chlorine Residual (mg/L)
1.6
1.4
1.2
1
0.8
0.6
0.4
0.2
0
1-Jan-98
30-Jun-98
27-Dec-98
25-Jun-99
22-Dec-99

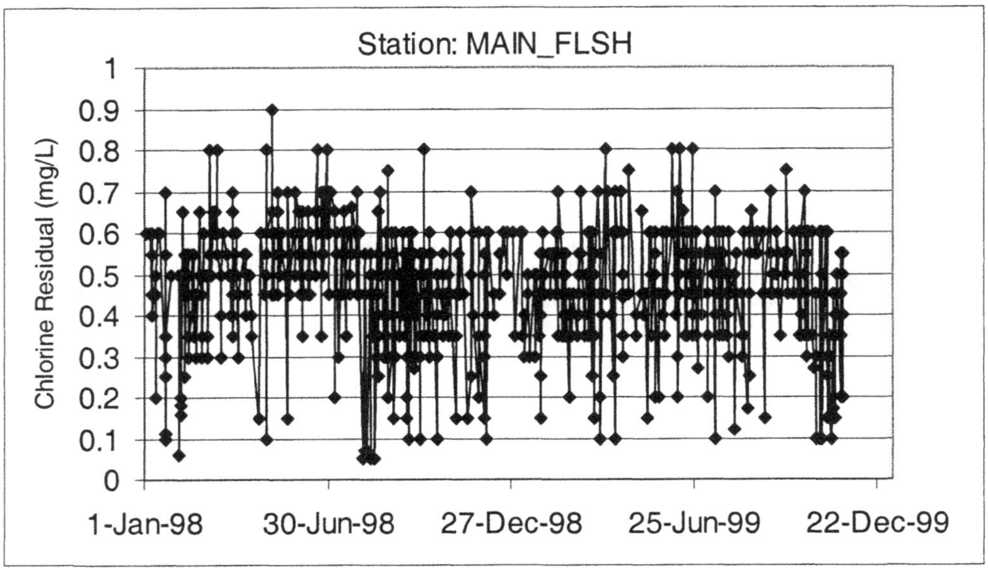
Station: MAIN_FLSH
Chlorine Residual (mg/L)
1
0.9
0.8
0.7
0.6
0.5
0.4
0.3
0.2
0.1
0
1-Jan-98
30-Jun-98
27-Dec-98
25-Jun-99
22-Dec-99

Plots of 1998-1999 Simulated Water Age and Chlorine Residual at Reservoirs

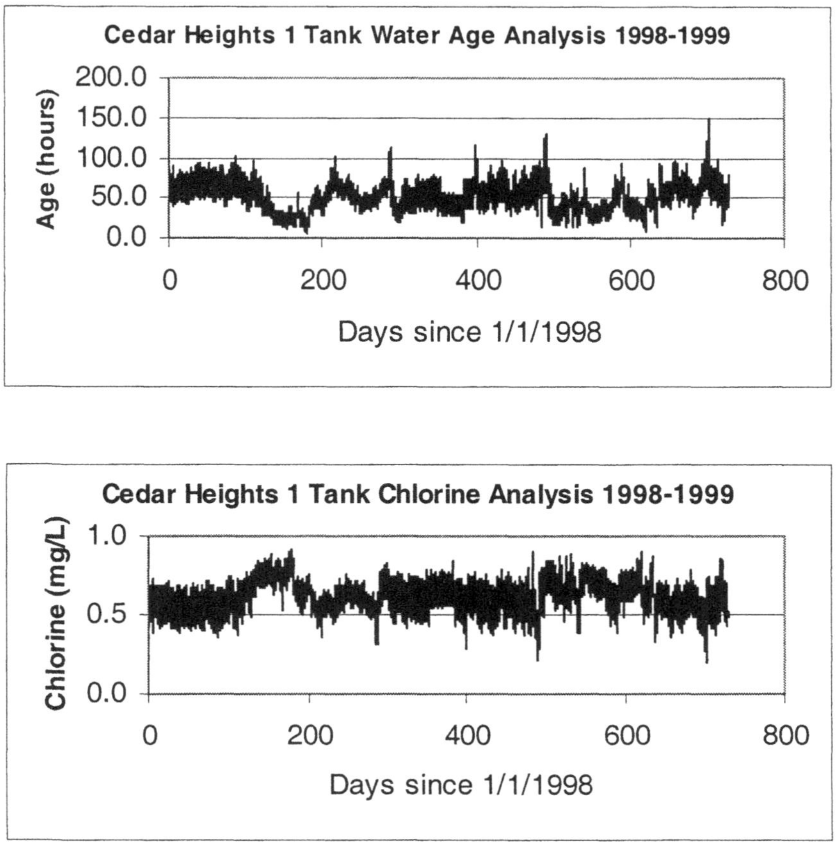

Figure A.1 Cedar Heights Tank #1

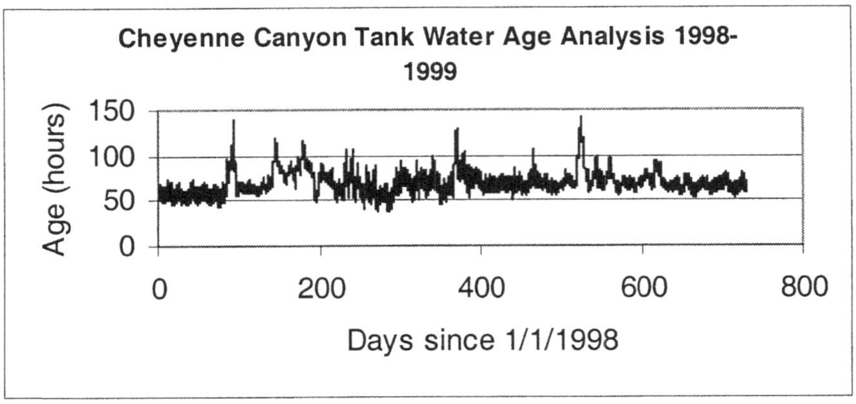

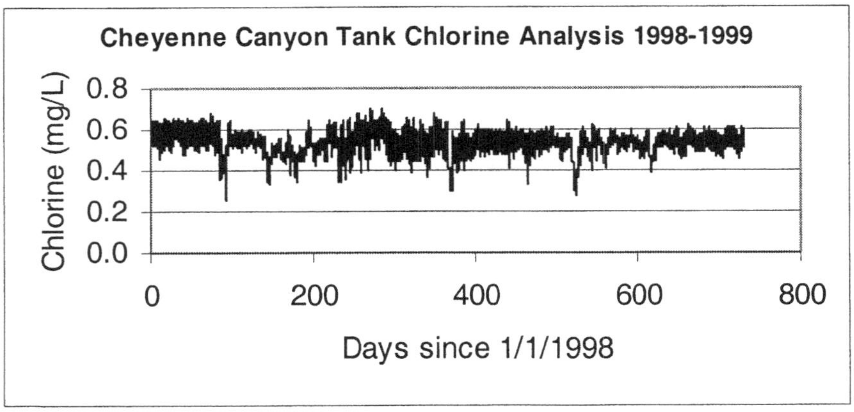

Figure A.2 Cheyenne Canyon Tank

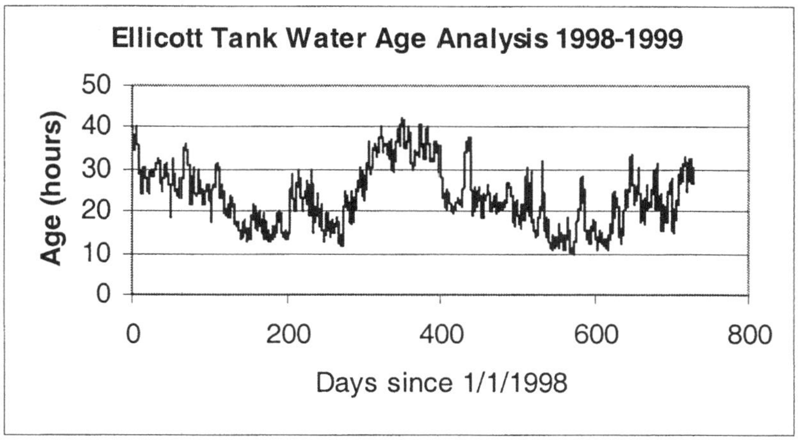

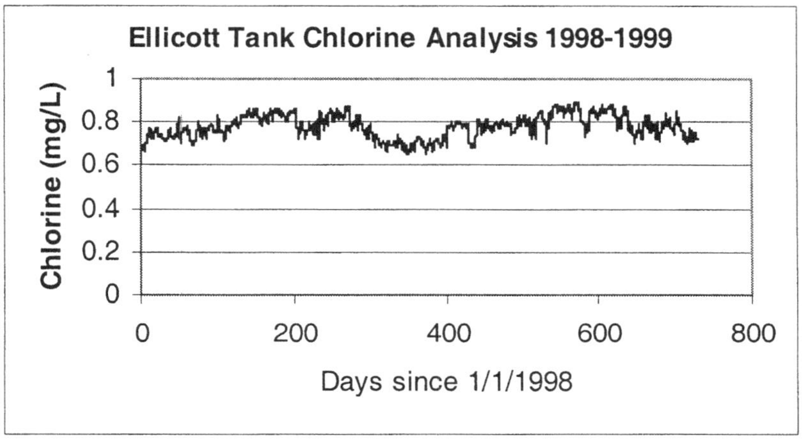

Figure A.3 Ellicott Tank

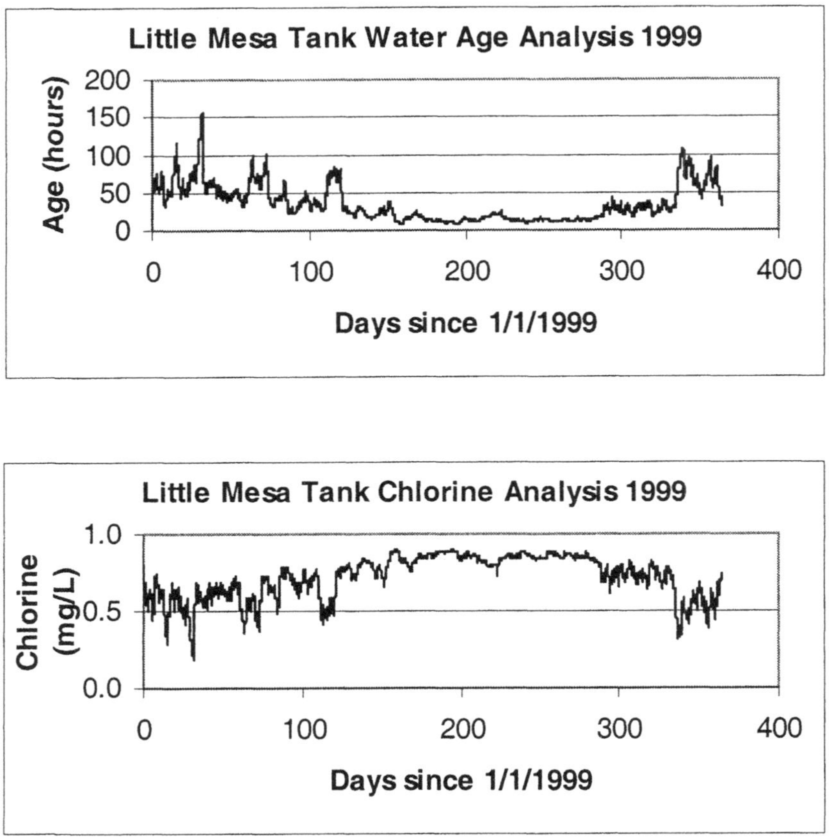

Figure A.4 Little Mesa Tank

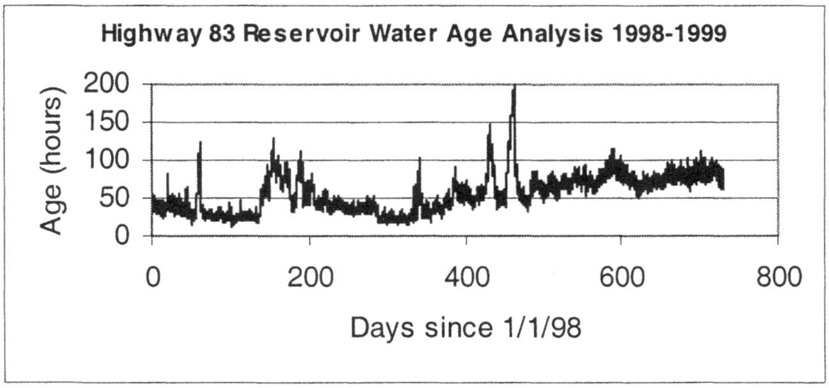

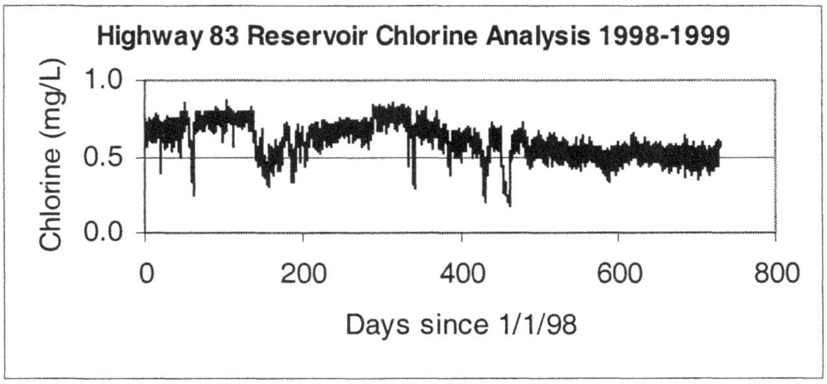

Figure A.5 Highway 83 Reservoir

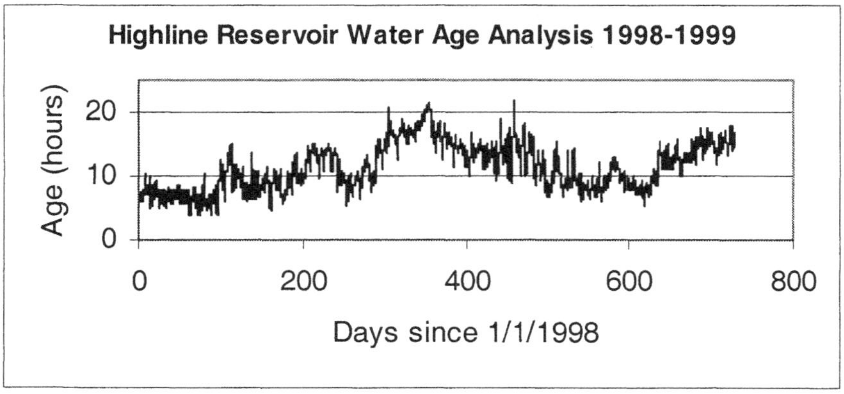

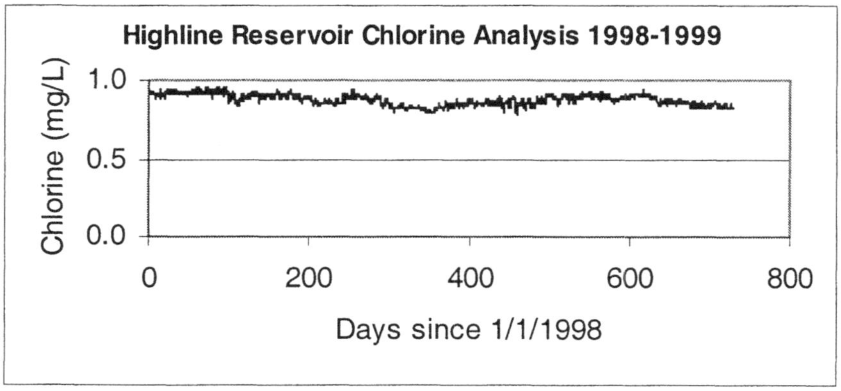

Figure A.6 Highline Reservoir

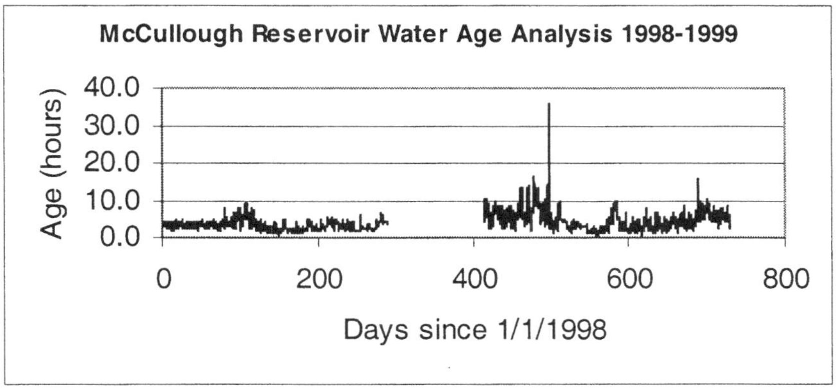

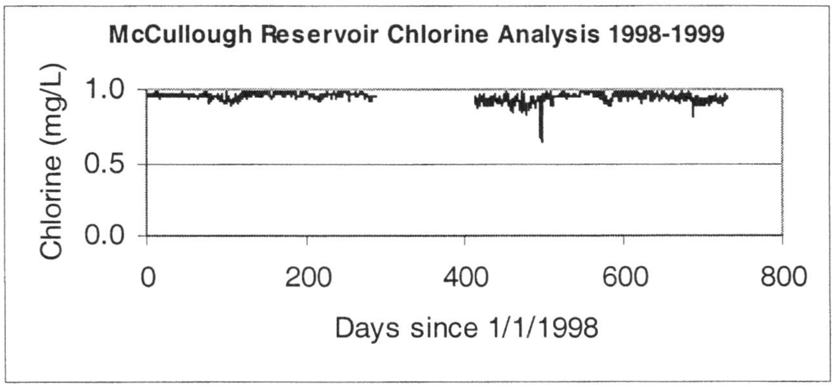

Figure A.7 McCullough Reservoir

APPENDIX B
JOB DESCRIPTION FOR OPERATIONS PLANNER AND SCHEDULER (OPS)

GENERAL STATEMENT OF DUTIES

Under general direction, prepares daily, weekly, monthly and annual operating plans for the water, wastewater, and non-potable systems. Plans contain production schedules for treatment plants and distribution/collection system control valve and pumping plant schedules. Coordinates system-wide operations and maintenance activities.

TYPICAL DUTIES AND RESPONSIBILITIES

1. Forecast water consumption and prepare operating plans which optimize the demand and cost of energy, commitment of labor, and quality of potable and non-potable water delivered to customers and wastewater collected, treated, and returned to the environment. Operating plans are based on the availability of water resources and predicted state of water/wastewater treatment, collection and distribution systems.
2. Monitor the operation of the water, wastewater, and non-potable systems and make real-time corrections to the operating plans based on unpredicted changes to the water system such as consumption fluctuations, equipment malfunction, pipe breaks, etc.
3. Coordinate, document, and assist in the planning of operations and maintenance activities on the utility systems. Maintain the clearance request system.
4. Provide technical support and training to System Operations to enable optimum operations of the utility systems.
5. Coordinate with other divisions in the preparation of long-range plans. Provide input and historical utility operations optimization information and define relationships to utility planning, engineering, and construction.
6. Coordinate with CSU Electric Department in the development of strategies for optimum hydro-generation and load management in the water, wastewater, and non-potable systems.
7. Communicate with and educate Water Resource Department Staff on the relationships between operations and maintenance activities and system optimization.
8. Participate in SCADA and other control systems development processes including needs assessments, design, implementation, testing, and operational assessment.
9. Perform other duties as required to maintain the reliability of the water, wastewater, and non-potable utility systems.

EDUCATION AND EXPERIENCE

Bachelor's degree in Engineering, Economics, Business or fifteen years of management/ supervisory experience in water or wastewater system operations.

Ten years of experience in utility system operations; utility system modeling experience is desirable.

SKILLS AND ABILITIES

- Experience in automated utility system operations
- Ability to apply and analyze utility system simulation models
- Ability to communicate effectively with technical, non-technical, management and operations/maintenance personnel
- Ability to work effectively in a team
- Ability to work in a complex environment and in periods of high stress
- Ability to create complex and detailed operating plans

APPENDIX C
FUNCTIONAL SPECIFICATION
CONSUMPTION FORECASTER

PURPOSE

The consumption forecasting application uses historical information about water system flows and weather, together with a weather forecast, to predict water system consumption a few days into the future.

The EWQMS system scheduler uses the consumption forecast to develop an optimal operating plan for the next 24 hours. Water system operators track actual consumption against forecasted consumption as one means of assessing how well they are operating the water system to the plan.

Water system operators also use the consumption forecast to help develop longer-range (i.e., longer than 24 hours) operation plans, schedule planned maintenance, plan raw water collection system operations, and perform other operations planning activities.

OBJECTIVES

Once a day, at an operator-specified time, an hourly consumption forecast will be prepared for 48 hours into the future. The hourly consumption forecast consists of the following water system elements:

1. Total system consumption
2. Consumption in the major pressure zones (or aggregated zones)
3. Consumption for the major water supply customers
4. Consumption for areas supplied by pumping stations

At any point in time, the 48-hour forecast can be updated after receiving an operator-entered command to do so.

INITIAL IMPLEMENTATION

The initial implementation will use simple time series analysis techniques combined with similar-day forecasting techniques to generate the consumption forecast.

FUTURE IMPLEMENTATION

Future versions of the forecasting application may use a neural-net forecasting technique or other advanced forecasting techniques. However, before considering implementing advanced forecasting techniques, CSU will undertake the following activities:

1. Perform preliminary data analysis as described below
2. Accumulate a long-term (two years) continuous historical database of the key consumption and weather data and forecasting parameters

3. Research successful applications of the technique at other water utilities

FUNCTIONAL DESCRIPTION

Overview

Once a day, at an operator-specified time, produce an hourly consumption forecast for 48 hours into the future. Develop the hourly consumption forecast for the following water system elements:

1. Total system consumption
2. Consumption in the major pressure zones (or aggregated zones)
4. Consumption for the major water supply customers, if applicable

Provide the capability for viewing actual consumption values together with forecasted consumption values in a trend format.

Similar-Day Searching and Review

1. Provide the capability to perform similar-day parameter searches on the historical consumption database to identify several (operator-selectable number) similar-day candidates within the historical record
2. Provide hourly trends of the similar-day consumption along with the forecasted consumption trend

The OPS must be able to review and adjust any or all forecasted consumption values prior to posting the final forecasted values for use by other applications.

At any point in time, update the 48-hour forecast after receiving an operator-entered command to do so.

Preliminary Analysis Needs and Objectives

Prior to implementing a consumption forecasting application, a thorough evaluation of water consumption patterns is required to identify the key parameters affecting consumption.

Objectives for the preliminary data analysis include:

1. Assemble a historical data record of consumption and potential driving parameters for subsequent forecasting model testing and tuning
2. Give operators and the OPS a better understanding of consumption patterns
3. Decide what level of detail (i.e., total system vs. major pressure zone vs. sub zone) is required for the consumption forecast

Data Required for Preliminary Analysis

At a minimum, the most recent full year of data is used for the preliminary data analysis. Several recent years covering a variety of overall climatic conditions is helpful.

Hourly and Daily Consumption Data

1. Consumption data is most often calculated from historical flow and tank volume information
2. Minimum requirement is total daily consumption for the entire service area
3. If possible, break total daily system consumption into consumption in each major pressure zone
4. If possible, develop hourly consumption values for the system as a whole and for each major pressure zone

Daily Weather Data

1. Total precipitation averaged over the entire service area
2. Maximum and minimum temperature averaged over the entire service area
3. Average wind speed over the service area
4. Maximum and minimum relative humidity over the service area
5. Any other weather data that may affect consumption for the utility

Wastewater Treatment Plant Inflow

1. CSU found wastewater treatment plant flow information to be helpful in establishing the indoor component of consumption – this is particularly true of combined water/wastewater utilities where the water and wastewater service areas are the same

Evapo-Transpiration Requirements

1. Some utilities have found the approximate evapo-transpiration requirements of typical residential, commercial and recreational lawns and gardens are well correlated to consumption
2. Evapo-Transpiration requirement is the daily total water requirements of typical residential and commercial lawns and gardens – this information can be estimated from weather data using standard formulas
3. Irrigation-Requirement is the daily evapo-transpiration requirement minus total daily rainfall (i.e., a typical lawn and garden's daily water deficit that needs to be supplemented by irrigation)

Items to Look For in the Preliminary Analysis

1. Establish the most important factors affecting consumption
2. Establish seasonal consumption patterns
3. Establish typical hourly consumption flow patterns for various seasons and days of the week
4. Establish the percent consumption concentration of each major pressure zone.
5. If possible see if there are distinct differences in seasonal, daily and hourly flow patterns between the major pressure zones

6. Determine if consumption seems to have distinct and easily separable indoor and outdoor use components
7. When water and wastewater service areas overlap, and sufficient flow data is available, determine if wastewater treatment plant flows are correlated to indoor consumption. If possible, separate the Infiltration and Inflow components of wastewater treatment plant flows from indoor water use.

Data Outputs and Output Interface

1. For the system as a whole and each consumption zone, generate an hourly consumption forecast extending two days (48 hours) into the future
2. The daily consumption values will be the daily average flow in units of million gallons per day (MGD)
3. The hourly consumption values will be hourly average flows in units of MGD – the hourly value time stamps will indicate the starting time of the hourly average flow period
4. As a minimum, the forecasting application will post updated forecast values to other applications via a Microsoft Excel spreadsheet file or as text data in a comma separated value file

Data Inputs and Input Interface

Final selection of the forecasting application inputs depends on the final solution technique and cannot be fully specified at this time. In general, the application will require the following inputs:

1. Historical water system data, assumed to be available via structured query language (SQL) query from an open database connectivity (ODBC) compatible database
   a. Historical record of actual consumption for the total system and for all of the main pressure zones
   b. Historical record of wastewater treatment plant flows, if applicable
   c. Historical record of weather data

Weather Forecast

A daily weather forecast is required for the current day and three days into the future including, but not limited to:

1. Expected maximum and minimum temperature
2. Expected precipitation
3. Expected cloudiness index or, alternatively, expected total solar radiation
4. Expected wind speed
5. Expected relative humidity

In the initial implementation, the OPS will manually enter and update the weather forecast information.

OPS Supervision and Control

1. Selected presentation of input data
2. Manual edits to the consumption forecast values
3. Command to submit a new forecast for posting to other applications
4. Forecasting model tuning parameter maintenance

Human-Machine Interface (HMI)

This section does describe the overall functionality and requirements for the consumption forecasting application HMI.

Forecast Projection Trend Displays

1. Provide trend displays to show the total system consumption forecast and also the individual consumption zone forecasts
2. On each trend, show the historical values for actual consumption, together with the forecasted values – the trend display should show a minimum of seven days historical data along with two days of forecasted data

Actual Versus Forecast Trend Displays

1. Provide trend displays that show actual versus forecasted consumption. Provide displays for total system consumption and also the individual consumption zones. Provide separate displays for hourly and daily forecasted values.
2. On each trend, show actual versus forecasted data as two separate trend lines. Actual consumption data will be consumption compiled within the water system information database. Forecasted data will be the most recent forecasted values (i.e., the forecasted values calculated one day prior to actual events).

Wastewater Treatment Plant Flow Display

1. Provide a trend display to show daily values of total wastewater treatment plant influent, estimated base flow, and estimated infiltration flow.

Similar Day Evaluation Display

1. Provide a display to permit the OPS to request a similar day search and display the results.
2. Permit the OPS to select various search parameters. Parameters may include, but not be limited to:
   a. Number of similar days to return
   b. Which forecast day to search for (e.g., current day, one day ahead, two days ahead)

Display the search results when they return. The results display may include, but not be limited to:

1. Date of similar day, along with total consumption, estimated base consumption and estimated irrigation consumption
2. Hourly consumption trends for each similar day plotted together on the same trend – also plot an hourly consumption trend based on the typical standard profile that best matches the search criteria
   a. Allow the OPS to over-ride the typical standard hourly profile with a selected typical day profile

<u>Forecast Posting Display</u>

1. Provide a display allowing the OPS to review the forecasting application's latest forecast, make final manual edits to the forecast, and to submit the updated forecast for posting to other applications
2. Forecast posting display functions may include, but not be limited to:
   a. Forecast day (i.e., current day, one day in the future, and two days in the future)
   b. For each forecast day show:
      1) Forecast base, irrigation, and total consumption
      2) Hourly forecast in trend format
   c. For each forecast day allow the OPS to modify the following:
      1) Base or irrigation consumption forecast values
      2) Hourly forecast values
      3) Provide a command to submit the final, edited forecast for posting to other applications

<u>Forecasting Application Tuning Parameter Maintenance Displays</u>

1. The exact nature of the application tuning parameter maintenance displays depends on the final solution technique – in general, however, provide displays that are clear, self-explanatory and easy to use

NOTES ON SOLUTION TECHNIQUE

This functional specification does not prescribe a specific solution technique for the consumption forecasting application. Most likely, the application will use a similar day forecasting and/or simple time series analysis techniques. One approach would be to first generate a forecast using the basic time series analysis, and then use a similar day search to provide a reasonability check on the forecast. Alternatively, the application could first do a similar day search and use the time series analysis forecast as a reasonability check. Note that the exact solution technique will depend on the results of the preliminary data analysis.

APPENDIX D
FUNCTIONAL SPECIFICATION
ECONOMIC PUMP/VALVE CONTROLLER

PURPOSE

The SCADA system Economic Pump/Valve Controller (EPVC) provides the means to control water system components (e.g., pumps and valves) to reduce energy cost and maintain high quality water. This specification defines the requirement to retrieve setpoints from an economic pump/valve setpoint file, and automatically control pumps and valves based on the time varying setpoint values. This specification is for the EPVC only and does not include a System Scheduler.

OBJECTIVES

The System Scheduler – Pump/Valve Optimizer will develop time varying pump/valve control setpoints (one minute resolution) for each controllable pump and valve in the water system. These setpoints will be generated to maintain high water quality and reduce the cost of energy required to meet daily water system consumption.

The EPVC will access the pump/valve setpoint tables every minute, compare the control variable to the setpoints, and send start/stop commands to two-state pumps. For variable speed pumps and modulating valves, the EPVC will read the pump/valve setpoint, compare it to the previous period's setpoint and download it to the RTU if there is a change from the last period. The EPVC will also check for pump/valve inhibits that will block downloading any commands or setpoints to the RTUs.

INITIAL IMPLEMENTATION

The EPVC will be initially implemented to support optimized control of distribution system pumps and valves. Raw water and non-potable pumping control will be implemented in a later phase of the project. The EPVC will be designed to permit optimized control of all controllable pumps and valves in the water system.

FUTURE IMPLEMENTATION

The EPVC is written to support initial and future applications of the EWQMS. The only changes in future applications would be to add more pumps and valves to support system expansion or renovation.

FUNCTIONAL DESCRIPTION

Overview

The Colorado Springs water system contains 180 booster pumps, 60 controllable PRVs and 25 flow control valves (FCVs) that will be controlled automatically or manually by the SCADA system.

The EPVC provides a hierarchy of control capabilities for the water system pumps and valves:

1. Local Control at the Site (both pumps and valves)
2. RTU control mode (pumps only) via existing automatic control programs and graphic displays
3. Manual control mode (valves only) via existing graphic displays
4. Operator Manual Control from the EPVC (pumps only)
5. Automatic Control from the EPVC (both pumps and valves)
6. Economic Control (Econ 1 or Econ 2) from the EPVC (both pumps and valves)

A lower numbered control mode precludes any higher numbered mode. Thus RTU control mode precludes EPVC Operator, Auto, Econ 1, and Econ 2 modes.

The EPVC module performs two main functions. For each two-state pump under automatic control, it compares the controlling variable to the on/off setpoint assigned to the pump, and initiates the required pump actions. For modulating valves and variable speed pumps, the EPVC will read the setpoint and, if changed from the previous period, download the setpoint to the RTU.

The EPVC task will wait a programmable length of time after system initialization before executing, allowing the system to stabilize. The task will execute every minute; this time period is programmable. The task will make a single pass through the entire list of pumps and valves every time it executes.

While this task can control all pumps or valves automatically, a pump or valve can be taken out of automatic control at the master station or at the remote site.

As previously defined, Inhibits are alarms that prevent pump start or valve control. Inhibits are defined on Automatic Pump and Valve Control Displays, and are logically OR'd with the result appearing in the inhibit column of the same displays. The occurrence of one inhibit alarm will block pump start or valve control. Pumps can be manually stopped in an alarm inhibit condition.

There exists a global "economic/auto" flag, which controls the Auto and Econ modes system-wide. When a global setpoint switchover is made, the master control status for each pump/valve in the system is changed. When the system is in Economic Mode, individual pumps/valves can be changed between the automatic and economic setpoint files through the Auto Pump Control and Automatic Valve Control Displays. However, individual pumps/valves cannot be changed from automatic to economic mode if the system is in automatic mode.

The system will usually be running with the Economic Setpoints, which are maintained by the System Scheduler software, or by the OPS. These Economic Setpoints represent the highest degree of optimization. There will be two economic setpoint tables for each pump/valve in the system. The economic setpoint tables will have a setpoint for each minute of the day or a total of 1440, setpoints. The control software will use the Economic Setpoints when the system is in the economic mode. Each pump or valve can be toggled between two economic setpoint tables – Econ1 or Econ2. In Econ1 mode the setpoints are manually created by the OPS. In Econ2 mode the setpoint tables are automatically updated by the System Scheduler software.

The Auto setpoints are constant time invariant values, which are manually entered by the Operator.

The Econ2 setpoint tables must be accessible for automatic file transfer from the System Scheduler, which will reside on another computer on the SCADA network.

Implement the system for controlling 180 two-state pumps, 85 modulation valves, and 10 variable speed pumps. The EPVC Software will be easily expandable for new pumps, valves, and control variables.

Definitions

"Called for" – When a remote unit (RTU) has received a 'pump on' command from the master station, it is said that the pump has been "called for." A "called for" can be issued or removed only when the switch of the RTU is in the remote or auto position. This is true for both automatic and manual mode.

Inhibit – An inhibit occurs when there is an alarm that should stop a pump or "inhibit" a pump from starting. When an inhibit alarm occurs, the master station will remove the "called for" from all affected pumps. The EPVC will remove the "called for" at any auto or econ mode pump affected by the inhibit. An inhibit will also prevent downloading a setpoint to a modulating valve or variable speed pump.

Local/Remote (L/R) or hand/off/auto (HOA) switch – Each pump or valve in the system has a L/R or HOA position switch at the remote site. The switch is read by the RTU to specify whether the pump can be controlled by local manual controls at the equipment or remotely via RTU or EPVC. This switch setting is telemetered to the master station. Pumps and valves in Local or Hand mode cannot be controlled by the EPVC.

RTU/EPVC mode for pumps – Each pump currently has an Auto/Manual mode selection. When the EPVC is installed, this mode selection will become an RTU/EPVC selection. In RTU mode, the pump will be controlled via the existing automatic control programs in the RTU and/or the existing graphic displays. In EPVC mode, the equipment will be controlled via the EPVC logic and EPVC control displays.

EPVC/Manual mode for valves – Each valve has an existing Auto/Manual mode selection. When the EPVC is installed, this mode selection will become an EPVC/Manual selection. In manual mode, the operators will manually open and close the valves via the existing graphic displays. In EPVC mode, the existing automatic control programs in the RTU will be active, but will get their flow and pressure setpoints from the EPVC instead of the existing graphic displays.

EPVC control modes:

1. Operator Mode from EPVC (two-state pumps only) – The operator manually commands pumps to start and stop from the EPVC pump control display.
2. Auto Mode from EPVC – For two-state pumps, the operator enters constant (not time-varying) pump start/stop level setpoints. For variable speed pumps, the operator

manually starts and stops pumps, but enters a constant flow or level setpoint. For valves, the operator enters constant flow or pressure setpoints.

3. Econ 1 Mode from EPVC – The operator enters a pattern of time varying pump start/stop level setpoints or valve flow or pressure setpoints. The EPVC automatically updates setpoints as dictated by the operator-entered time schedule.
4. Econ 2 Mode from EPVC – the System Scheduler, Pump/Valve Optimizer automatically generates time-varying pump start/stop level setpoints or valve flow or pressure setpoints.

System Scheduler – Software programs that develop optimized control setpoints for pumps and valves in the water system. The System Scheduler minimizes energy cost for system operations under the operating parameters generated by the Water Quality Analyzer and Water Source Analyzer. The System Scheduler consists of an optimizer and water system simulator programs.

Two-State Pump Control Logic

The diagram in Figure D.1 represents the EPVC two-state pump control logic.

1. For a two-state pump to be controllable via the EPVC, it must be placed in Remote Mode at the site, and placed in an EPVC mode via the RTU/EPVC mode flag (formerly auto/manual flag)
2. For a pump to turn on and off automatically in EPVC control, it must be selected in Auto, Econ 1 or Econ 2 mode. In these modes, each pump is assigned start/stop setpoints, which are compared to the control variable. When the control variable falls to or below the ON setpoint, the pump will turn on. When the control variable equals or exceeds the OFF setpoint, the pump will turn off.

Based on the measured and selected control variable and monitored pump status and inhibit inputs, the control strategy sets or resets the control outputs that start/stop individual pumps. The pump scan and tag attributes are also checked prior to issuing a command or setpoint. The control variable scan attribute is also checked, and a control is inhibited if the control variable is not in scan.

Any time an alarm condition (inhibit) in the field occurs and a two-state pump is stopped by interlocks in the RTU, the EPVC, upon recognition of this event, will issue a stop command to synchronize the EPVC and RTU with the field condition. When the alarm condition clears, the program will automatically restart the two-state pumps if setpoint conditions are satisfied.

The final step prior to issuing a start command to a two-state pump is to check the status of the pump stagger start timers. The stagger start times are used to prevent hydraulic and electric surges associated with the simultaneous starting of the pumps at a pump station or over cycling pumps. The stagger start timers are started after the master station issues a start command. No other pumps can be started at this facility until the group stagger time expires and the individual pump stagger time expires. The stagger start timers for pumps are adjustable from the Auto Pump Control Display (Figure D.4).

Variable Speed Pump and Valve Control Logic

Variable Speed Pump Control

1. The diagram in Figure D.2 illustrates control logic at the master station for variable speed pumps.
2. The pump must be in Remote mode and EPVC mode for the EPVC to have control of the pump.
3. The pump must be in Auto, Econ 1, or Econ 2 modes for the EPVC to automatically turn pumps on and off and/or adjust setpoints based on operator entries or a time schedule.
4. In Auto, Econ 1, or Econ 2 modes, the EPVC task will read the pump run status and setpoint tables, and only start a pump if the run status is set from the Auto or Econ run status tables. Prior to sending the start command, the EPVC task will download the setpoint. If the pump is already running and the setpoint changes, the EPVC task will download a new setpoint.

Modulating Valve Control

1. The diagram in Figure D.3 illustrates control logic at the master station for modulating valves.
2. The valve must be in Remote mode and EPVC mode for the EPVC to have control of the valve.
3. In EPVC mode, the EPVC task will read the setpoint tables and only download new setpoints to the remote site if the setpoints have changed, the valve is not tagged, an inhibit is not set, and the control variable is in scan.

Manual Control Via EPVC Displays

Manual control from the EPVC displays applies to pumps only. Note that manual control of valves is still possible from the existing graphics displays.

Two-State Pumps

1. Manual control of two-state pumps from the EPVC is called "operator" mode.
2. When a two-state pump's L/R switch is in the Remote position, the pump in EPVC mode, and the pump is in EPVC "operator" mode, the control variables are no longer compared to setpoints. Instead, the operator can manually start and stop pumps from the EPVC displays as described in the display descriptions below.
3. When the Operator uses a display to issue a command, the supervisory control software will check control requirements and generate the appropriate message for transmission to the RTU.

Variable Speed Pumps

1. "Operator" mode does not apply to variable speed pumps.

2. When variable speed pumps are in Remote Mode, EPVC Mode, and are in Auto mode within the EPVC. The operator can manually start and stop pumps and manually download flow/level setpoints to the pump.
3. The supervisory control software will check control requirements and generate the appropriate message for downloading to the RTU.

Displays

Several displays are needed to provide access to information relevant to the automatic control subsystem. The OPS and Operators use displays to examine the current status of the individual pumps/valves, to view pump/valve setpoints and parameters, and to view and alter system automatic control parameters.

The following displays are required:

1. Automatic Pump Control
2. Automatic Valve Control

<u>Automatic Pump Control</u>

Figure D.4 is the Automatic Pump Control Display. The mode column shows which master station control mode the pump is in. In AUTO mode, each two-state pump is controlled by the control variable value. The AUTO setpoints are fixed and changeable by the OPS or Operator. Each pump is turned on when the control variable falls to or below the ON level and turned off when the control variable rises to or above the OFF setpoint.

The Econ1 and Econ2 setpoints function in the same manner. Variable speed pumps have setpoints downloaded when the Operator, OPS, or System Scheduler changes the setpoints. In manual mode, the Operator directly controls two-speed pumps.

The following is a description of the Automatic Pump Control Display. There is one display for each pump station in the water system; a description for each column follows.

1. PUMP CONTROL PULL-DOWN MENU – Options for changing the control mode, manually controlling a two-state or variable speed pump, or downloading a setpoint. To manually download a setpoint to a variable speed pump, change the Auto setpoints, select download setpoint, and then Execute or Cancel. To manually control a two-state or variable speed pump; select the pump, change the mode to Operator, select Start or Stop, then select Execute (Cancel if necessary).
2. STATION – The pump station name is automatically indicated when the display is selected.
3. GROUP STAGGER TIME SET – Time entered for the pump station Group Stagger Timer.
4. PUMP STA TIME REMAIN – Time remaining on the Group Stagger Timer.
5. SYSTEM PUMP MODE – Indicates if the pumps in the system are in Automatic or Economic Modes. System can be transferred between modes via the push buttons below.
6. PUMP SELECT – Pump that has been selected for control or mode change. Selecting pump will activate pump control pull-down menu.

7. PUMP – Name of the pump.
8. RUN STATUS – Current status of the pump from the database.
9. CALLED FOR – Two-state pump has been "called for" from Auto, Econ, or Operator Modes.
10. INHIBIT – Indicates if a control inhibit condition exists. The inhibit condition is the logical "OR" of the inhibit alarms listed below.
11. TAG – Indicates if the pump is tagged out of service. Tag status is automatically updated from the database.
12. LOCAL – (Yes) Pump is in local control and cannot be controlled by the master station.
13. (No) Pump is controllable from the master station.
14. RTU MODE – Current mode of RTU control. EPVC mode means control by the EPVC software program at the master station. RTU mode means control by RTU program at the site.
15. EPVC MODE – Current control mode of the pump. Constant Speed Pump mode is changed by first selecting a pump then selecting Operator, Auto, Econ1, or Econ2 from the pull-down menu. Variable Speed Pump mode is changed by first selecting Pump, then selecting Auto, ECON1, ECON2. There is no Operator Mode for variable speed pumps.
16. ECON SETPOINTS – On and Off setpoints that are used to control the pump when it is in the Economic Mode. Setpoints are automatically passed from the System Scheduler setpoint files in the Econ2 mode. For the Econ1 mode, provide a pull-down menu such that Economic Setpoints can be manually entered for each pump by identifying a time period and the On/Off setpoints for that period. Displays will be provided to show all setpoints for the day for both Econ1 and Econ2 modes. Note that variable speed pumps only have one setpoint and a Run command for the two Economic modes.
17. AUTO SETPOINTS – Constant On and Off setpoints that are used to control the pump when it is in Auto mode. Note that Variable Speed Pumps only have one setpoint and a Run command in the Auto mode. The OPS or the Operator manually enters Setpoints.
18. PUMP STAG TIME – Individual pump stagger timer value entered by OPS or operator.
19. PUMP STAG REMAIN – Time remaining on the individual pump stagger timer.
20. CONTROL VARIABLES – The point name, current value, and engineering units for the control variables.
21. INHIBITS – Inhibit point names for each pump inhibit. Inhibits are set and changed here.

Automatic Valve Control Display

The Automatic Valve Control Display is similar to the Automatic Pump Control Display and is illustrated in Figure D.5. There will be valve control displays to accommodate all remote controllable valves in the system. All flow control valves will be on flow control valve display. Remotely controlled PRVs will be in PRV displays. A description for each column follows:

1. VALVE CONTROL PULL-DOWN MENU – Options for changing the control mode or manually downloading a setpoint or deadband to a valve. To download a setpoint or deadband, first select the value, then select the function (setpoint or deadband download), and finally select Execute (Cancel if necessary).
2. FLOW OR PRV – Flow control valve or PRV display.
3. SYSTEM VALVE MODE – Indicates if the valves in the system are in automatic or economic modes. System can be transferred between modes via the push buttons below.
4. VALVE SELECT – Valve that has been selected for control or mode change. Selecting a valve will activate valve control pull-down menu.
5. VALVE NAME – Name of the valve.
6. VALVE CONTROL – Valve can be pressure (P) or flow (F) control. Econ mode will always be flow control.
7. UP PRESSURE – Current upstream pressure at the valve.
8. DOWN PRESSURE – Current downstream pressure at the valve.
9. FLOW – Current flow through the valve.
10. INHIBIT – Indicates if a control inhibit condition exists. The inhibit condition is the logical "OR" of the inhibit alarms listed below.
11. TAG – Indicates if the valve is tagged out of service. Tag status is automatically updated from the database.
12. RTU MODE – In manual it means manual control of valve from graphic displays. In EPVC mode it means automatic or economic setpoint control loaded by EPVC application.
13. MODE – Same as Pump Control Display except there is not an Operator mode.
14. AUTO SETPOINTS – Setpoints entered by the operator or OPS.
15. AUTO SETPOINT UNITS – Setpoint units in up pressure, down pressure flow.
16. ECON SETPOINTS – Setpoints that are updated from the valve setpoint tables. Setpoints can be manually updated from the economic setpoint tables (Econ1) or automatically from the System Scheduler (Econ2). Provide a pull-down menu such that Economic Setpoints can be manually entered for each valve (Econ1) by identifying a time period and the setpoint for that period. Displays will be provided to show all setpoints for the day.
17. ALARM DB – An alarm will be generated when the measured FLOW or pressure, compared to the SETPOINT FLOW or pressure, exceeds the deadband defined in this column.
18. ALARM DB TIME – Time in minutes that deadband must be exceeded to generate an alarm.
19. SETPOINT DEVIATION ALARM – Indicates there is a valve setpoint alarm.
20. CONTROL VARIABLES – The point name, current value, and engineering units for the control variables.
21. INHIBITS – Alarms that will inhibit economic setpoint control.

System Scheduler

The interface between the AI SCADA System and the System Scheduler software is shown in Figure D.6. As shown in Figure D.6, the interface between the EPVC and the System Scheduler is via pump and valve Economic Setpoint Tables transferred via middleware.

The System Scheduler software will reside on an Application Server on the SCADA LAN. In the Econ2 mode, the ON/OFF pump setpoints and valve setpoints will be developed by the System Scheduler on a pump or valve basis with a resolution of one (1) minute. In Econ1 and Econ2 modes, economic setpoints will be developed for each controllable pump and valve for every minute of the day.

Economic Setpoints will initially be developed by the OPS or Operator in the Econ1 mode. Ultimately, new Economic Setpoints will be generated automatically by the System Scheduler software in the Econ2 mode.

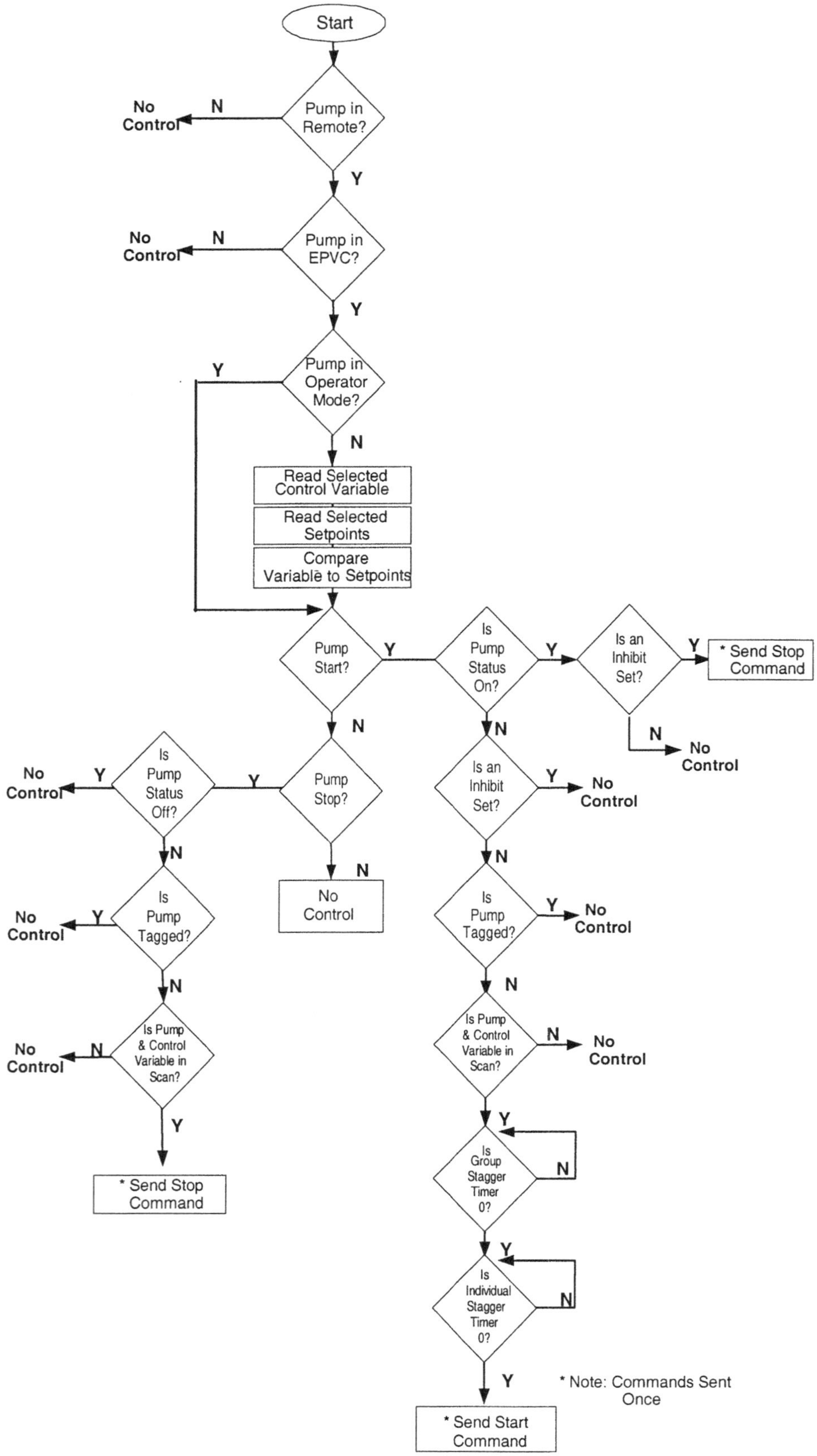

Figure D.1 Two-state Pump Control Logic

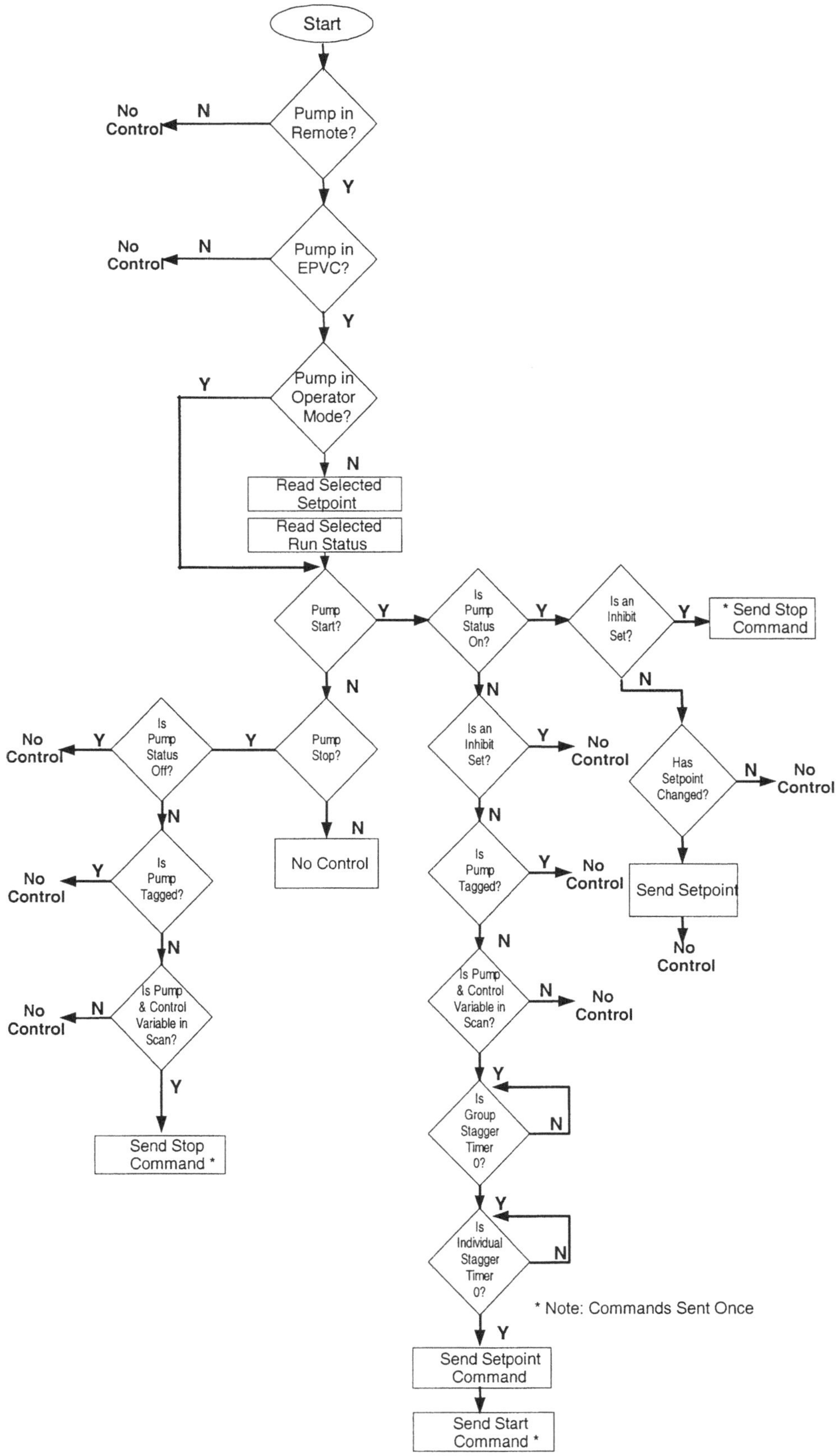

Figure D.2 Variable Speed Pump Control Logic

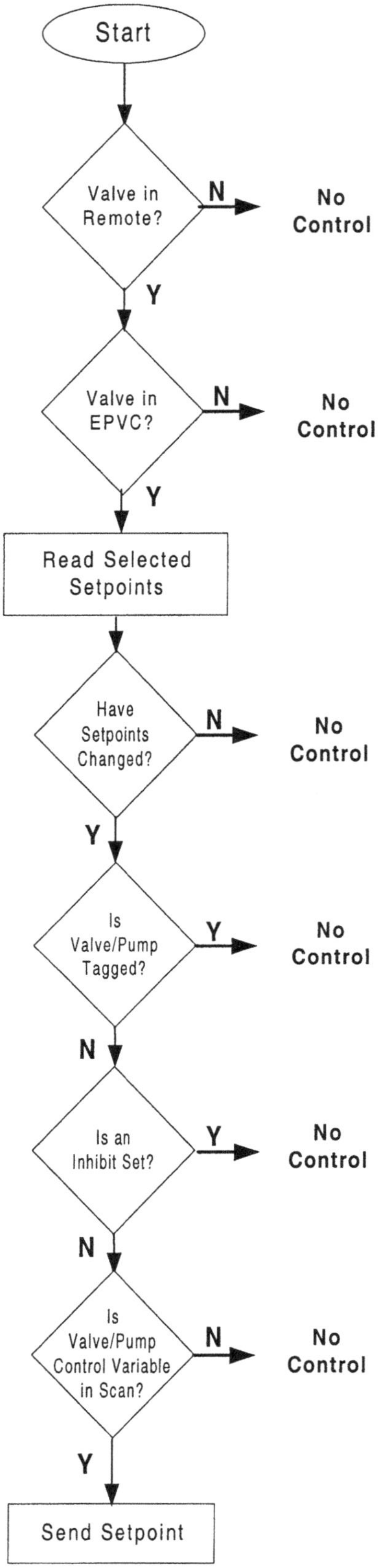

Figure D.3 Modulating Valve Control Logic

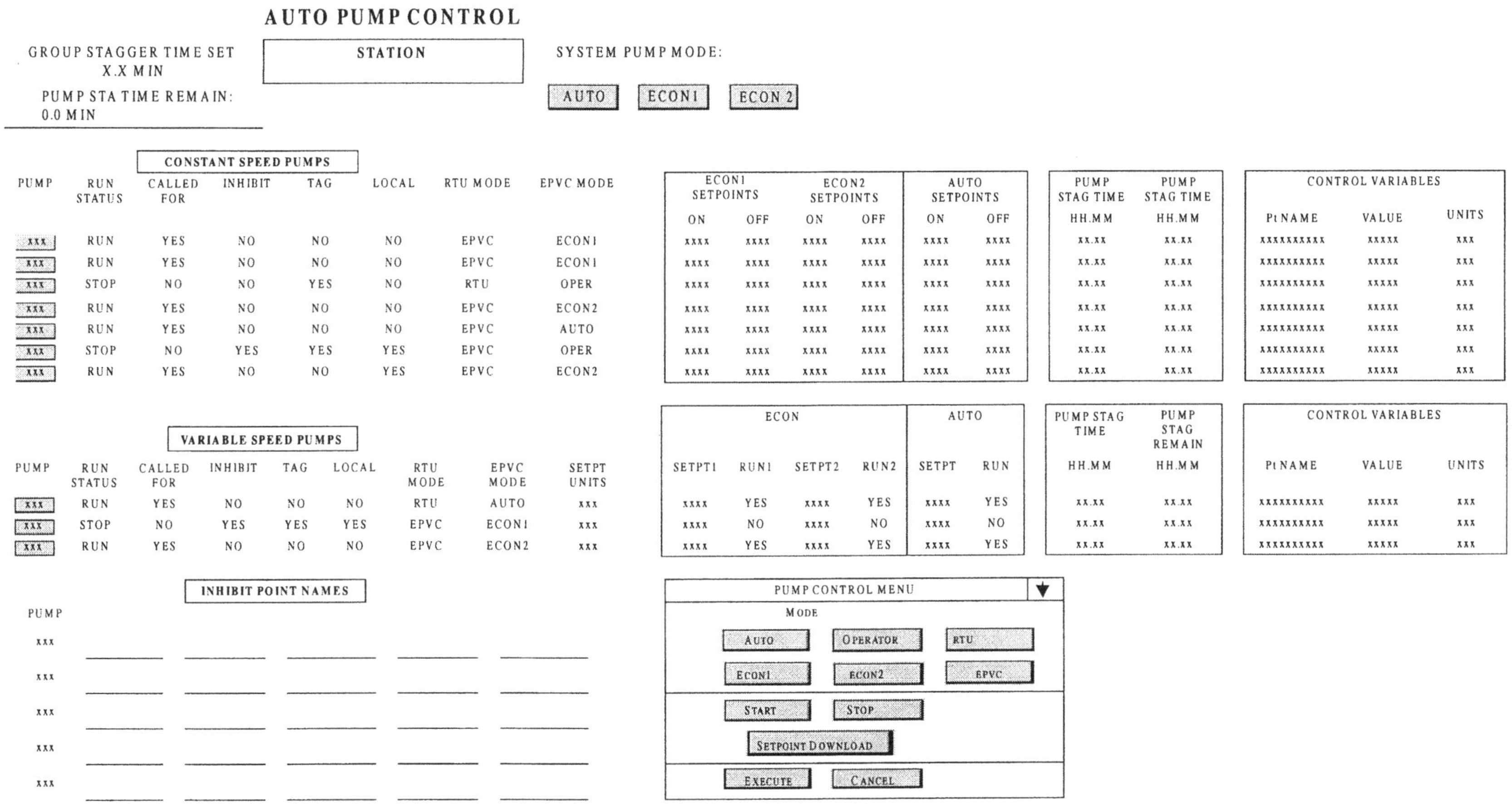

Figure D.4 Automatic Pump Control Display

(An 11 x 17 version of this figure is available at the back of the report.

AUTOMATIC VALVE CONT ROL

FLOW OR PRV DISPLAY

SYSTEM VALVE MODE: ECON2

[Manual] [Econ1] [Econ2]

Valve Name	Valve Control	Up Pressure (PSI)	Down Pressure (PSI)	Flow (MGD)	Inhibit	Tag	RTU Mode	EPVC Mode	Setpoints: Auto Setpoint	Setpoints: Auto Setpoint Units	Econ Setpoints: ECON1	Econ Setpoints: ECON2	Control Deadband
xxxxxxxxxxxxxxxxx	P	xxx.x	xxx.x	xx.xx	NO	NO	MANUAL	ECON1	xxx.x	UP	xxx.x	xx.xx	x.xx
xxxxxxxxxxxxxxxxx	F	xxx.x	xxx.x	xx.xx	YES	YES	EPVC	AUTO	xxx.x	DP	xxx.x	xx.xx	x.xx
xxxxxxxxxxxxxxxxx	P	xxx.x	xxx.x	xx.xx	NO	NO	EPVC	ECON2	xxx.x	F	xxx.x	xx.xx	x.xx
xxxxxxxxxxxxxxxxx	P	xxx.x	xxx.x	xx.xx	NO	NO	EPVC	ECON2	xxx.x	F	xxx.x	xx.xx	x.xx
xxxxxxxxxxxxxxxxx	F	xxx.x	xxx.x	xx.xx	NO	NO	EPVC	ECON2	xxx.x	F	xxx.x	xx.xx	x.xx

VALVE TUNING

Valve Name	Alarm DB (MGD)	Alarm DB Time (Min)	Setpoint Deviation Alarm	Control Variable: Name	Control Variable: Value	Control Variable: Units
xxxxxxxxxxxxxxxxx	x.xx	xx.x	Normal	xxxxxxxxxx	xxxxxx	xxx
xxxxxxxxxxxxxxxxx	x.xx	xx.x	Normal	xxxxxxxxxx	xxxxxx	xxx
xxxxxxxxxxxxxxxxx	x.xx	xx.x	Normal	xxxxxxxxxx	xxxxxx	xxx
xxxxxxxxxxxxxxxxx	x.xx	xx.x	Normal	xxxxxxxxxx	xxxxxx	xxx
xxxxxxxxxxxxxxxxx	x.xx	xx.x	Alarm	xxxxxxxxxx	xxxxxx	xxx

VALVE CONTROL MENU

Mode

[Econ1] [Econ2] [Manual] EPVC

[Deadband Download]

[Setpoint Download]

[Execute] [Cancel]

INHIBITS

Valve Name					
xxxxxxxxxxxxxxx					
xxxxxxxxxxxxxxx					
xxxxxxxxxxxxxxx					
xxxxxxxxxxxxxxx					
xxxxxxxxxxxxxxx					

Figure D.5 Automatic Valve Control Display

(An 11 x 17 version of this figure is available at the back of the report.

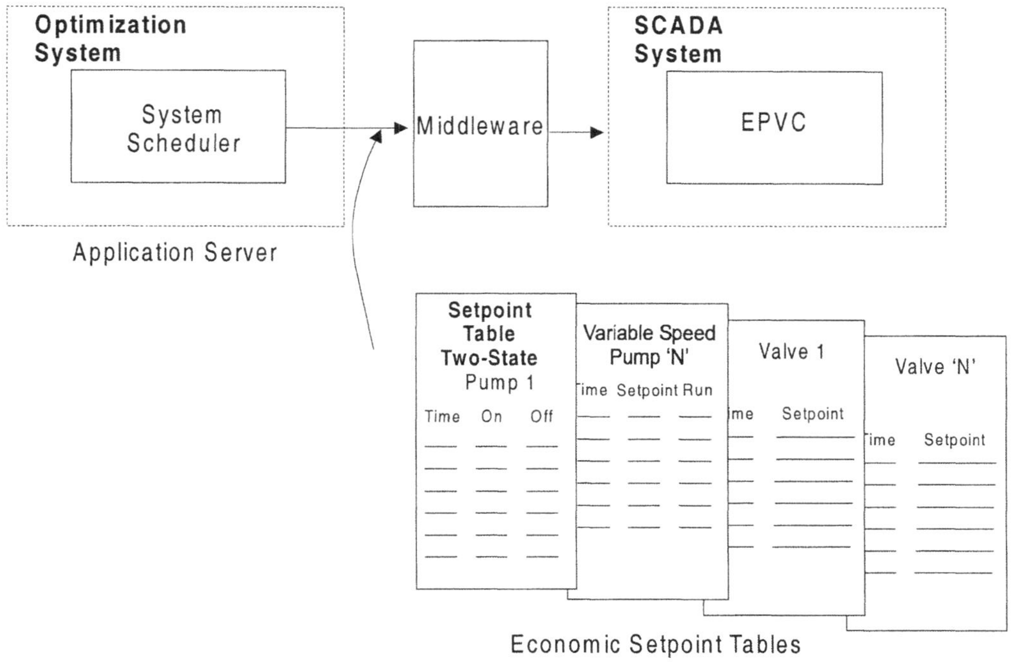

Figure D.6 System Scheduler Interface to Economic Pump/Valve Controller

APPENDIX E
FUNCTIONAL SPECIFICATION
WATER SYSTEM SIMULATOR

PURPOSE

The Water System Simulator is used by the OPS to study water system operation under different scenarios. It allows the OPS to verify that a proposed pump/valve control schedule satisfies constraints and identifies constraints that may be violated.

OBJECTIVES

The Simulator uses a system model and operating schedule information such as flow from treatment plants, pump and valve schedules, consumption, and energy costs. For a specified time period (e.g., one day), it simulates system operation in suitable time increments and determines tank levels, flow through valves, and operating costs. The Simulator supports an object-oriented graphical user interface (GUI) that displays the system diagram and allows the user to modify schedules and settings. Schedules and results are displayed in a graphic form on the system diagram. Constraint violations such as tank levels are flagged. Results from the simulation are stored in the repository for further analysis.

INITIAL IMPLEMENTATION

Initially the Simulator will use a mass balance model.

FUTURE IMPLEMENTATION

In future phases, it is anticipated that a detailed hydraulic model will be used to replace or assist in tuning the mass balance models.

FUNCTIONAL DESCRIPTION

Overview

The purpose of the Simulator is to provide the OPS with a tool to simulate the operation of the water system for a given time period. To accomplish this, the Simulator utilizes historical or projected consumption information about various water system areas, or zones, in conjunction with pump and valve operating schedules. For each minute during the Simulation time period, the simulator calculates flow through each PRV, calculates flow from active pumps and water treatment plants, and subtracts consumption in each water consumption area. At the end of the simulation time period the Water System Simulator displays the result of the simulation, including the optimum pump schedule, and stores this information for further analysis.

Simulator Inputs

1. Time period (e.g., one day), time to be simulated
2. Time steps (e.g., one minute or 15 minutes)
3. Input/Output time increment (e.g., 15 or 30 minutes)
4. Starting reservoir levels, water treatment plants (WTPs) and Tesla schedules
5. Consumption for the time period for each Input/Output time increment
6. Energy costs for each Input/Output time increment
7. Maintenance schedules (pump/valve tag outs)

Simulator Outputs

Calculate at the end of each Input/Output time increment (calculations will be performed using the time step specified):

1. Reservoir level
2. Flow at key points
3. Energy cost by pump
4. Water age

Graphical Outputs

Update data repository so that the user has access to graphical displays (trends over the specified period) of:

1. Water consumption
2. Pump status
3. Reservoir levels (with violation flags)
4. KWH consumption in each Input/Output time increment energy cost

Object Definitions and Attributes

Object	Attributes (Units of Measure)
Water Service Area (Zone)	Elevation (feet)
Reservoir	Elevation (feet) Capacity (kgals) Set Overflow (feet or %full) Set Fire Storage (feet or %full) Minimum Water Quality Level (feet or %full)
Tank	Elevation (feet) Capacity (kgals) Overflow Fire Storage
Hydro Plant	Elevation (feet) Flow Range (gpm) Set Range (gpm) Capacity (mgd) Kwh per Kgal (kwh)
WTP	Elevation (feet)

Object	Attributes (Units of Measure)
	Flow Range (gpm) Set Range (gpm) Capacity (mgd) Kwh per Kgal (kwh)
Pump	Elevation (feet) Rated GPM (gpm) Actual GPM (gpm) Kwh per Kgal (kwh) Pump status (Open/Closed)
PRV	Elevation (feet) Zone Assignment In-Out (Zone to and from) Max Flow (gpm) Diameter (inches)
FCV	Elevation (feet) Zone Assignment In-Out (Zone to and from) Max Flow (gpm) Diameter (inches)
Consumer Consumption Point	Kgals per interval (kgal)

Specific Objects Defined for CSU

	Water Service Area (Zone):
1.	Briargate
2.	Highline
3.	Lowline
4.	Northfield
5.	Templeton

	Storage Tanks:
1.	Austin Bluffs ASTB3
2.	AFA #4 AFA43
3.	Bear Creek 2 – Upper Skyway
4.	Briargate BRIG3
5.	Broadmoor #1 & 5 BRM13, BRM53
6.	Broadmoor #3 & 3A, & 4
7.	Cedar Heights #1 CDR12
8.	Cedar Heights #2 CDR22
9.	Cedar Heights #3 CDR33
10.	Cheyenne Canyon CHEY3
11.	Ellicott ELEC3
12.	FVA Terminal
13.	Highline HIGH1
14.	Highway 83 HWY83
15.	Little Mesa LMES3
16.	Lower Neal Ranch NEAL2
17.	Upper Neal Ranch UPNL3
18.	McCullough MCUL1
19.	Northgate NOGT2
20.	Pine Valley PVLY1
21.	Research RSLX2

Storage Tanks:
22. Sunrise SUNR2
23. Templeton TMPL2
24. Upper Woodmen Valley UWVL2
25. Wilson WLSN2
26. Woodmen Valley WDMN3

WTPs (Sources):
1. Fountain Valley Authority
2. McCullough Treatment Plant
3. Mesa Treatment Plant
4. Pine Valley Treatment Plant

PRVs/FCV:
1. 90WV4
2. ABMR
3. ACAP4
4. ACGL4
5. ACML5
6. ACWN4
7. APCR4
8. BFV-3
9. BLFR4
10. CDR12
11. CHY84
12. CMBV4
13. CNVN4
14. CSCP4
15. DRPW
16. ELEC3
17. ELECTRIC-VALVE
18. FLYR5
19. FRAC4
20. FTCE6
21. FTCW6
22. HCNC4
23. HCVJ4
24. JSVO
25. KCEL5
26. LMES3
27. LXOR
28. MBUN4
29. MCUL6
30. MDGG
31. NFHL4
32. NFLD5
33. NVPV
34. PETE6
35. PINE-VALLEY-CONTROL-VALVE
36. PVLY6
37. PWAP4
38. RSLX213
39. RSLX215
40. RSLX217
41. RVSC4

PRVs/FCV:
42. SGHY
43. SMRS
44. SRBV
45. TJCP4
46. WLSN4
47. WN254
48. WNAA4
49. WNUN4

Pumps:
1. Bear Creek 2 BEAR2 a. BEAR201 b. BEAR202
2. Broadmoor 4 BRM42 a. BRM4201 b. BRM4202
3. Cedar Heights 1 CDR12 a. CDR1201 b. CDR1202
4. Cedar Heights 2 CDR22 a. CDR2201 b. CDR2202
5. Gold Camp/21st St. GC212 a. GC21201 b. GC21202 c. GC21203
6. Howbert/27th HB272 a. HB27201 b. HB27202
7. Lower Sykway SKYL2 (backup) a. SKYL201 b. SKYL202 c. SKYL203
8. Neal Ranch NEAL2 a. NEAL201 b. NEAL202
9. Northgate NOGA2 a. NOGA201 b. NOGA202 c. NOGA203
10. Northgate NOGT2 (backup) a. NOGT201 b. NOGT202
11. Oro Blanco OROB2 (backup) a. OROB201 b. OROB202
12. Paisley PSHW2 a. PSHW201 b. PSHW202 c. PSHW203

Pumps:
13. Pine Valley – AFA PVLY2 a. PVLY201 b. PVLY202 c. PVLY203 d. PVLY204 e. PVLY205 f. PVLY206
14. Point of Pines POFP2 a. POFP201 b. POFP202
15. Research RSLX2 a. RSLX201 b. RSLX202 c. RSLX203 d. RSLX204
16. Rio Grande/8th St. RG8S2 a. RG8S201 b. RG8S202
17. Roxbury ROX22 a. ROX2201 b. ROX2202 c. ROX2203
18. Roxbury ROXB2 (backup) a. ROXB201 b. ROXB202
19. Sunrise SUNR2 a. SUNR201 b. SUNR202 c. SUNR203
20. Templeton TMPL2 a. TMPL201 b. TMPL202 c. TMPL203 d. TMPL204 e. TMPL205
21. University Park UNPK2 a. UNPK201 b. UNPK202 c. UNPK203 d. UNPK204
22. Upper Woodman UWVL2 a. UWVL201 b. UWVL202 c. UWVL203
23. Westmeadow WMDL2 a. WMDL201 b. WMDL202 c. WMDL203 d. WMDL204
24. Wilson WLSN2 a. WLSN201 b. WLSN202 c. WLSN203
25. Wintwood WWCP2

Pumps:
a. WWCP201 b. WWCP202

Reservoirs:
1. Rampart
2. Tesla

Consumption Points:
1. Air Force Academy
2. Bear Creek-2 – Upper Skyway
3. Briargate
4. Broadmoor-4
5. Broadmoor-5
6. Cedar Heights-1
7. Cedar Heights-2
8. Cedar Heights-3
9. Fort Carson
10. Highline - East
11. Highline – SW
12. Highline East
13. Highway 83
14. Lowline
15. Mountain Shadows
16. Neal Ranch
17. Northfield
18. NorthGate
19. Paisley
20. Peterson Air Force Base
21. Point of Pines
22. Reduced Briargate
23. Sunrise
24. Templeton
25. University Park
26. Upper Neal
27. Upper Woodman
28. Wilson
29. Woodman
30. Woodman Valley

APPENDIX F
FUNCTIONAL SPECIFICATION
OPTIMUM PUMP SCHEDULER

PURPOSE

The Optimum Pump Scheduler is used by the OPS to develop a water system operating plan (i.e., schedule of treatment plant flows, pump operation, and valve flow) such that operating constraints are met, and energy costs are minimized, and benefit of Tesla generation is maximized.

OBJECTIVES

The optimum pump scheduler uses a system model, information such as inflow to treatment plants, pump and valve availability, consumption, energy costs, and constraints from water quality and water rights. For a specified time period (e.g., one day) it determines a schedule (i.e., treatment plant flows, valve flow settings, and pump status). The schedule is expressed in suitable increments of time (e.g., every 30 minutes for the specified time period).

The schedule is such that:

- Operating constraints such as tank levels are not violated
- Energy costs incurred under specified energy cost structures are minimized
- The generation schedule for Tesla Hydro-generator is scheduled in a way to maximum its value based on input provided by electric operations

The Water System Simulator (Appendix E) will be used to display results for approval or modification.

INITIAL IMPLEMENTATION

The optimization of pump schedules is a mathematical challenge. It will be prudent in Phase 1 to accept a simple and inexpensive solution to implementation that produces sub-optimal results. However, it is required that these results be acceptable in a practical sense and able to be easily modified by the OPS or an experienced operator. In the initial implementation a simplified mass-balance model will be used.

FUTURE IMPLEMENTATION

In future phases, it is anticipated that more sophisticated algorithms, including hydraulic models, will be developed.

FUNCTIONAL DESCRIPTION

Overview

The purpose of the optimum pump scheduler is to provide OPS with a tool to study optimum operating strategies for water system pump, valves and hydro-generator.

The principal application is in developing the scheduling component of the daily operating plan, namely, develop optimum pump schedules based on energy costs *and* develop hydro-generator schedules based "value at different times of day" as provided by electric operations. Note there are effectively two objectives that need to be optimized.

Additionally, the OPS can use the optimizer to:

- Modify and re-evaluate hydro-generator and treatment plant schedules based on review of hydro-generator operating schedule by OPS and electric operations
- Evaluate effects of maintenance and outages
- Evaluate effect of different energy cost structures

Pump/Valve Optimization

Initially, booster pumps in the water system operate under Time-Of-Use (TOU) rates. Thus the energy optimization problem is to confine pumping to periods of low energy cost. In the simplified "mass-balance" model of the water system we can imagine that water is delivered to tanks through flow control valves and via booster pumps. The tanks supply consumption points. In turn flow control valves should be scheduled to maintain tank levels that allow pumps to run in low-cost periods.

Tesla-Hydro Optimization

The Tesla hydro-electric plant is a unique feature of the CSU system. The plant supplies water to treatment plants (primarily McCullough WTP) and generates electricity. This electric energy is the cheapest resource for CSU. As evaluated in Chapter 2, there is significant benefit in maximizing the energy generated.

Tesla generation is constrained by the amount of water that can be taken from the after-bay by the water treatment plants. Thus low water consumption constrains the electric generation. It is also desired that WTP flow remain reasonably constant.

The optimization of Tesla hydro-electric generation is therefore considered to be an iterative process (Figure 3.11). The approach is envisioned as follows:

1. CSU electric provides information regarding the "value" or "benefit" of Tesla generation for each hour of the day.
2. OPS uses the optimizer, specified in this appendix, to maximize an index composed of Tesla generation (or flow) multiplied by the "value." In other words, Tesla generation is highest when its value is perceived to be highest. The optimizer results must, of course, satisfy level constraints on the forebay and after-bay, and constraints on flow through the hydro-generator.
3. The OPS reviews the resulting generation schedule with CSU electric operations. If appropriate, the "value" definitions are modified and the study repeated.

Optimizer Inputs

1. Time period (e.g., one day) to be optimized
2. Time steps (e.g., one minute or 15 minutes)
3. Input/Output time increment (e.g., 15 or 30 minutes)
4. Starting reservoir level
5. Initial Pump/Valve schedule (status) for the time period for each Input/Output time increment
6. Maintenance schedules and pre-schedules
7. Consumption for the time period for each Input/Output time increment
8. Energy costs for each Input/Output time increment
9. Utility performance criteria (Tesla Hydro-generator value)

Optimizer Outputs

1. Produce a least energy cost schedule. This schedule specifies flow from the treatment plants, pump status, and valve flow, in each time increment. Practically, this schedule will be as close to optimal as can be achieved within modeling and computing resource constraints.
2. Provide schedules in text form for review by user.
3. Display results through the Simulator (see Specifications Appendix E).

Objective Definitions and Attributes

The optimizer will utilize the following objects and their associated attributes. The following exceptions will be allowed:

1. Parallel flow control valves may be combined into one valve
2. Reservoirs, water treatment plants, and tanks may be modeled by a single object type
3. Nodes can be modeled as tanks with the requirement that level remain the same at all times
4. Tesla hydro-generator may be modeled as a pump or a valve
5. Consumption points may be combined
6. Consumption points may be allocated to suitable tanks

Object	Attributes (Units of Measure)
Water Service Area (Zone)	Name
Reservoir	Elevation (feet) Capacity (kgals) Set Overflow (feet or %full) Set Fire Storage (feet or %full) Minimum Water Quality Level (feet or %full) (Future)
Tank	Elevation (feet) Capacity (kgals) Overflow Fire Storage
Hydro Plant	Elevation (feet) Flow Range (gpm) Set Range (gpm) Capacity (mgd) Kwh per Kgal (kwh)
WTP	Elevation (feet) Flow Range (gpm) Set Range (gpm) Capacity (mgd) Kwh per Kgal (kwh)
Pump	Elevation (feet) Rated GPM (gpm) Actual GPM (gpm) Kwh per Kgal (kwh) Pump status (Open/Closed)
PRV	Elevation (feet) Zone Assignment In-Out (Zone to and from) Max Flow (gpm) Diameter (inches)
FCV	Elevation (feet) Zone Assignment In-Out (Zone to and from) Max Flow (gpm) Diameter (inches)
Consumer Consumption Point	Kgals per interval (kgal)

Specific Objects Defined for CSU

Water Service Area (Zone):
1. Briargate
2. Highline
3. Lowline
4. Northfield
5. Templeton
Storage Tanks:
1. Austin Bluffs ASTB3
2. AFA #4 AFA43
3. Bear Creek 2 – Upper Skyway
4. Briargate BRIG3

	Storage Tanks:
5.	Broadmoor #1 & 5 BRM13, BRM53
6.	Broadmoor #3 & 3A, & 4
7.	Cedar Heights #1 CDR12
8.	Cedar Heights #2 CDR22
9.	Cedar Heights #3 CDR33
10.	Cheyenne Canyon CHEY3
11.	Ellicott ELEC3
12.	FVA Terminal
13.	Highline HIGH1
14.	Highway 83 HWY83
15.	Little Mesa LMES3
16.	Lower Neal Ranch NEAL2
17.	Upper Neal Ranch UPNL3
18.	McCullough MCUL1
19.	Northgate NOGT2
20.	Pine Valley PVLY1
21.	Research RSLX2
22.	Sunrise SUNR2
23.	Templeton TMPL2
24.	Upper Woodmen Valley UWVL2
25.	Wilson WLSN2
26.	Woodmen Valley WDMN3
	WTPs (Sources):
1.	Fountain Valley Authority
2.	McCullough Treatment Plant
3.	Mesa Treatment Plant
4.	Pine Valley Treatment Plant
PRVs/FCV:	
1.	90WV4
2.	ABMR
3.	ACAP4
4.	ACGL4
5.	ACML5
6.	ACWN4
7.	APCR4
8.	BFV-3
9.	BLFR4
10.	CDR12
11.	CHY84
12.	CMBV4
13.	CNVN4
14.	CSCP4
15.	DRPW
16.	ELEC3
17.	ELECTRIC-VALVE
18.	FLYR5
19.	FRAC4
20.	FTCE6
21.	FTCW6
22.	HCNC4
23.	HCVJ4
24.	JSVO

PRVs/FCV:
25. KCEL5
26. LMES3
27. LXOR
28. MBUN4
29. MCUL6
30. MDGG
31. NFHL4
32. NFLD5
33. NVPV
34. PETE6
35. PINE-VALLEY-CONTROL-VALVE
36. PVLY6
37. PWAP4
38. RSLX213
39. RSLX215
40. RSLX217
41. RVSC4
42. SGHY
43. SMRS
44. SRBV
45. TJCP4
46. WLSN4
47. WN254
48. WNAA4
49. WNUN4

Pumps:
1. Bear Creek 2 BEAR2 a. BEAR201 b. BEAR202
2. Broadmoor 4 BRM42 a. BRM4201 b. BRM4202
3. Cedar Heights 1 CDR12 a. CDR1201 b. CDR1202
4. Cedar Heights 2 CDR22 a. CDR2201 b. CDR2202
5. Gold Camp/21st St. GC212 a. GC21201 b. GC21202 c. GC21203
6. Howbert/27th HB272 a. HB27201 b. HB27202
7. Lower Sykway SKYL2 (backup) a. SKYL201 b. SKYL202 c. SKYL203
8. Neal Ranch NEAL2 a. NEAL201 b. NEAL202
9. Northgate NOGA2

Pumps:
a. NOGA201 b. NOGA202 c. NOGA203
10. Northgate NOGT2 (backup) a. NOGT201 b. NOGT202
11. Oro Blanco OROB2 (backup) a. OROB201 b. OROB202
12. Paisley PSHW2 a. PSHW201 b. PSHW202 c. PSHW203
13. Pine Valley – AFA PVLY2 a. PVLY201 b. PVLY202 c. PVLY203 d. PVLY204 e. PVLY205 f. PVLY206
14. Point of Pines POFP2 a. POFP201 b. POFP202
15. Research RSLX2 a. RSLX201 b. RSLX202 c. RSLX203 d. RSLX204
16. Rio Grande/8th St. RG8S2 a. RG8S201 b. RG8S202
17. Roxbury ROX22 a. ROX2201 b. ROX2202 c. ROX2203
18. Roxbury ROXB2 (backup) a. ROXB201 b. ROXB202
19. Sunrise SUNR2 a. SUNR201 b. SUNR202 c. SUNR203
20. Templeton TMPL2 a. TMPL201 b. TMPL202 c. TMPL203 d. TMPL204 e. TMPL205
21. University Park UNPK2 a. UNPK201 b. UNPK202 c. UNPK203 d. UNPK204
22. Upper Woodman UWVL2

Pumps:
a. UWVL201 b. UWVL202 c. UWVL203
23. Westmeadow WMDL2 a. WMDL201 b. WMDL202 c. WMDL203 d. WMDL204
24. Wilson WLSN2 a. WLSN201 b. WLSN202 c. WLSN203
25. Wintwood WWCP2 a. WWCP201 b. WWCP202

Reservoirs:
1. Rampart
2. Tesla

Consumption Points:
1. Air Force Academy
2. Bear Creek-2 – Upper Skyway
3. Briargate
4. Broadmoor-4
5. Broadmoor-5
6. Cedar Heights-1
7. Cedar Heights-2
8. Cedar Heights-3
9. Fort Carson
10. Highline - East
11. Highline – SW
12. Highline East
13. Highway 83
14. Lowline
15. Mountain Shadows
16. Neal Ranch
17. Northfield
18. NorthGate
19. Paisley
20. Peterson Air Force Base
21. Point of Pines
22. Reduced Briargate
23. Sunrise
24. Templeton
25. University Park
26. Upper Neal
27. Upper Woodman
28. Wilson
29. Woodman
30. Woodman Valley

APPENDIX G
FUNCTIONAL SPECIFICATION
WATER QUALITY ANALYZER

PURPOSE

Examines water quality monitoring data and related supporting data and determines whether any values are out of acceptable range.

OBJECTIVES

The water quality analyzer will accept the following general types of data that relate to water quality:

- Real-time water quality data
- Data accessed from a Laboratory Information Management System (LIMS)
- Real-time hydraulics data

The analyzer will perform various analyses, and the resulting information will be compared to baseline water quality values. If the resulting values are considered to be out of normal ranges then the OPS and/or operator is notified of potential water quality problems. The Water Quality Analyzer could also be used in conjunction with the System Scheduler to consider water quality objectives and/or constraints in setting the daily schedule.

INITIAL IMPLEMENTATION

Initially the analyzer will focus on a surrogate water quality indicator in the distribution system – water age. A water age calculation module has been integrated into the system simulator.

FUNCTIONAL DESCRIPTION

On a daily basis, the Water System Simulator will simulate behavior of the system over the next 24 hours and will determine the water age at each of the consumption nodes and tanks. The calculated water age will be compared by the Water Quality Analyzer to baseline values to determine whether the predicted water age is within acceptable limits. Two general baseline water age values will be used throughout the system corresponding to:

- Areas that are receiving water that was chlorinated only at the treatment plants
- Areas receiving water that has been re-chlorinated at booster stations in the distribution system

Typical initial baseline ages for these two areas are four days and seven days respectively.

FUTURE IMPLEMENTATION

In future phases of EWQMS, several upgrade paths are possible for the water quality analyzer. These options are outlined below:

- Model and assess predicted chlorine levels in the distribution system. In a manner similar to the daily water age calculations, a chlorine calculation module has been designed for use in the system simulator. On a daily basis, the system simulator would predict the chlorine residual at consumption nodes and tanks over the next 24-hour period. Chlorine residuals below a pre-set alarm value (e.g., 0.2 mg/L) would alert the operator that a potential problem could exist in terms of low chlorine values.
- For each on-line chlorine monitor in the distribution system, an alarm level will be identified. Typical alarm values are 0.2 mg/L. Real-time chlorine measurements will be compared to the alarm value and values of chlorine residual less than the alarm values will result in an alarm to the operator.
- Examination of non-real-time water quality data accessed from a LIMS. Information entered into a LIMS could be examined by the Water Quality Analyzer, compared to norms for the appropriate parameter, and integrated with other information available to the Analyzer in order to identify potential water quality problems.
- Examine other parameters. Other chemicals or water quality indicators could be examined by the analyzer through modeling, on-line monitors or examination of LIMS data.
- Integration of the Water Quality Analyzer into the System Scheduler is another option. Ultimately, the Water Quality Analyzer could be integrated with the System Scheduler so that water quality data and predicted (modeled) water quality values could directly influence the schedule for the next 24-hour period. Water quality objectives and constraints could be considered by the System Scheduler along with energy related factors to determine an "optimal" schedule for the next 24-hour period.

APPENDIX H
FUNCTIONAL SPECIFICATION
WATER SOURCE ANALYZER

PURPOSE

The Water Source Analyzer (WSA) application uses information about raw water system flows (historical), treatment costs, pumping rates/costs, storage value, local source value, hydro-generation potential value, together with water quality information, to optimize raw water transmission and WTP production scheduling a few days into the future. Figure H.1 illustrates the functional layout of the WSA. Figure H.2 depicts the use-case for generation of the WTP production schedule using the WSA. Benefits include reduced costs based on optimizing hydroelectric generation and pumping.

The EWQMS System Scheduler uses the Water Source Analyzer to develop an initial optimal operating plan for raw water transmission and WTP production targets. The EWQMS System Scheduler then accomplishes further optimization for hydroelectric generation and pumping load for the next 24 and 48 hours. Water system operators track actual raw water source transmission and WTP production against forecasted flows as a means of assessing how well they are operating the water system to the plan.

OBJECTIVES

The WSA provides an optimized schedule of raw water source transmission volumes for supply of WTPs. Additionally, a schedule of WTP production volumes is provided.

Once a day, at an operator-specified time, the WSA produces a daily raw water source and WTP production schedule is produced for 24 and 48 hours into the future. The following describes the WSA process:

1. OPS Input of zone and system-wide consumption forecast to the WSA
2. Determination of WTP Production Targets based on availability of raw water sources and yields
3. Determination of Raw Water Source supplying WTPs is based on priority (determined by optimization) and yields

At any point in time, the WSA updates the 24- and 48-hour forecasts after receiving an operator-entered command to do so.

INITIAL IMPLEMENTATION

The initial WSA implementation will use optimization techniques based on genetic algorithms to select an optimum solution for raw water sources supplying WTPs and an initial schedule for WTP production. Application design will include spreadsheets with an optimization plug-in and linkage to EWQMS applications databases through ODBC.

FUTURE IMPLEMENTATION

Future versions of the WSA may be improved to include additional water quality constraints, updated pumping efficiency/load information, and greater provision for linkage with hydro-optimization. Enhancements will include improved linkage with EWQMS application databases.

FUNCTIONAL DESCRIPTION

Overview

The objective of the WSA is to produce an Optimized Water Source Priority Schedule. Totalizing relative costs from the following cost categories generates the Water Source Cost/Value List:

- Input of relative costs for pumping of each raw water source
- Input of relative costs for treatment of each raw water source
- Input of relative costs (value) for hydro-generation from each raw water source
- Input of relative costs (value) for draw-down from local terminal storage for each raw water source
- Input of relative costs (value) for drawing from local sources for each raw water source

Constraints are applied for water quality, source capacity, source availability, and system response time. Constraints for optimization of water sources are established based on current operating practices and limitations of the Raw Water Transmission, Treatment and Distribution systems. An optimized water source schedule is then generated based on the relative costs provided in the Water Source Cost/Value List while in conformance with all established constraints, but resulting in the "lowest cost" alternative solution. The result of optimization is a listing of water sources with a take volume schedule for each source. This optimized schedule is referred to as the "Water Source Priority Schedule." Additionally, a "WTP Production Schedule" is generated based on the most current water system consumption forecast.

Once a day, at an operator-specified time, produce a daily "Water Source Priority Schedule" and "WTP Production Schedule" (optimized Water Source Priority Schedule) for 24 and 48 hours into the future. The OPS and operator may view actual raw water transmission and WTP production values together with forecasted values in a trend format. The OPS may review and adjust any or all forecasted values prior to posting the final forecasted values for use by other applications. At any point in time, update the 24- and 48-hour forecast after receiving an operator-entered command to do so.

Preliminary Analysis Needs and Objectives

Prior to implementing a Water Source Analyzer application, water utilities need to evaluate raw water transmission requirements and characteristics, and identify the key variables that affect scheduling of sources.

Objectives for the preliminary analysis include:

1. Assemble a historical data record of transmission volumes, available yields and driving parameters (i.e., water quality data such as fluoride, turbidity)
2. Give operators and OPS a better understanding of raw water transmission patterns
3. Describe and document transmission operating rules (driven by water rights, priorities and regulatory requirements) that affect scheduling of sources

At a minimum, the most recent full year of data is used for the preliminary data analysis. Recent years covering any unusual operating conditions/constraints would be helpful. Monthly raw water transmission volumes, available yields, and driving parameters (including water quality constraints) are necessary. Records of energy usage/generation, flows, and costs are required if hydroelectric power generation and/or pumping apply. Treatment cost information for each source is also required.

Based on raw water transmission rules/constraints, establish relative costs for each source that describe and influence source selection for optimization of water source transmission schedule. Modify relative costs to ensure rules/constraints are met.

Verify that Water Treatment Plant (WTP) production schedules are practicable and meet constraints for system capacity, flow limitations, and meet consumption forecast requirements.

Verify optimized transmission pumping costs with actual pumping data. Ensure that pumping schedules do not violate capacity limits.

Verify feasibility of transmission schedules with system capacity and available yields data.

Verify that storage constraints are met for optimized schedule of raw water transmission.

Data Outputs and Output Interface

For the system as a whole, generate a daily raw water transmission and WTP production forecast extending one day (24 hours) and two days (48 hours) into the future. The daily schedule will be the daily average flows in units of MGD. As a minimum, the Water Source Analyzer will post updated forecast schedules to other applications via an MS Excel spreadsheet file or as text data in a comma-delineated file.

Data Inputs and Input Interface

The WSA application inputs include a total daily water system consumption forecast. Water source and WTP availability, water source historical yields, static pump and system capacity information, and energy cost/production information are collected for use in the WSA. Relative cost tables are established, based on inputs of hydro-generation values, pumping cost, local source values, draw-down storage values, treatment costs, and miscellaneous cost adjustment values.

OPS Supervision and Control

The following are supervisor and control functions available to the OPS:

1. Data presentation as specified in the HMI paragraphs
2. Manual edits to all WSA inputs

3. Manual edits to the raw water transmission and WTP production forecasted values
4. Command to submit a new forecast for posting to other applications
5. Water Source Analyzer forecast tuning parameter maintenance

Human-Machine Interface (HMI)

WSA Forecast Projection Trend Displays

1. Provide trend displays to show the total and individual WTP production and raw water source transmission forecasts.
2. On each trend, show the historical values for actual WTP production and raw water source transmission together with the forecasted values. The trend display should show a minimum of seven days historical data along with two days of forecasted data.

WSA Actual versus Forecast Trend Displays

1. Provide trend displays that show actual versus forecasted WTP production and raw water source transmission. Provide displays for total WTP production and raw water source transmission.
2. On each trend, show actual versus forecasted data as two separate trend lines. Actual production/transmission data will be compiled within the water system information database. Forecasted data will be the most recent forecasted values.

WSA Application Tuning Parameter Maintenance Displays

1. Simple spreadsheet entry forms with minimal data requirements (utilizing direct query of databases wherever possible).
2. Reports allowing comparison of actual with forecasted flow data.

NOTES ON SOLUTION TECHNIQUE

This functional specification prescribes a specific solution technique for the optimization of raw water transmission and WTP production scheduling; however, many techniques are available for the optimization of scheduling. The CSU WSA application utilizes genetic algorithms provided by off-the-shelf spreadsheet software.

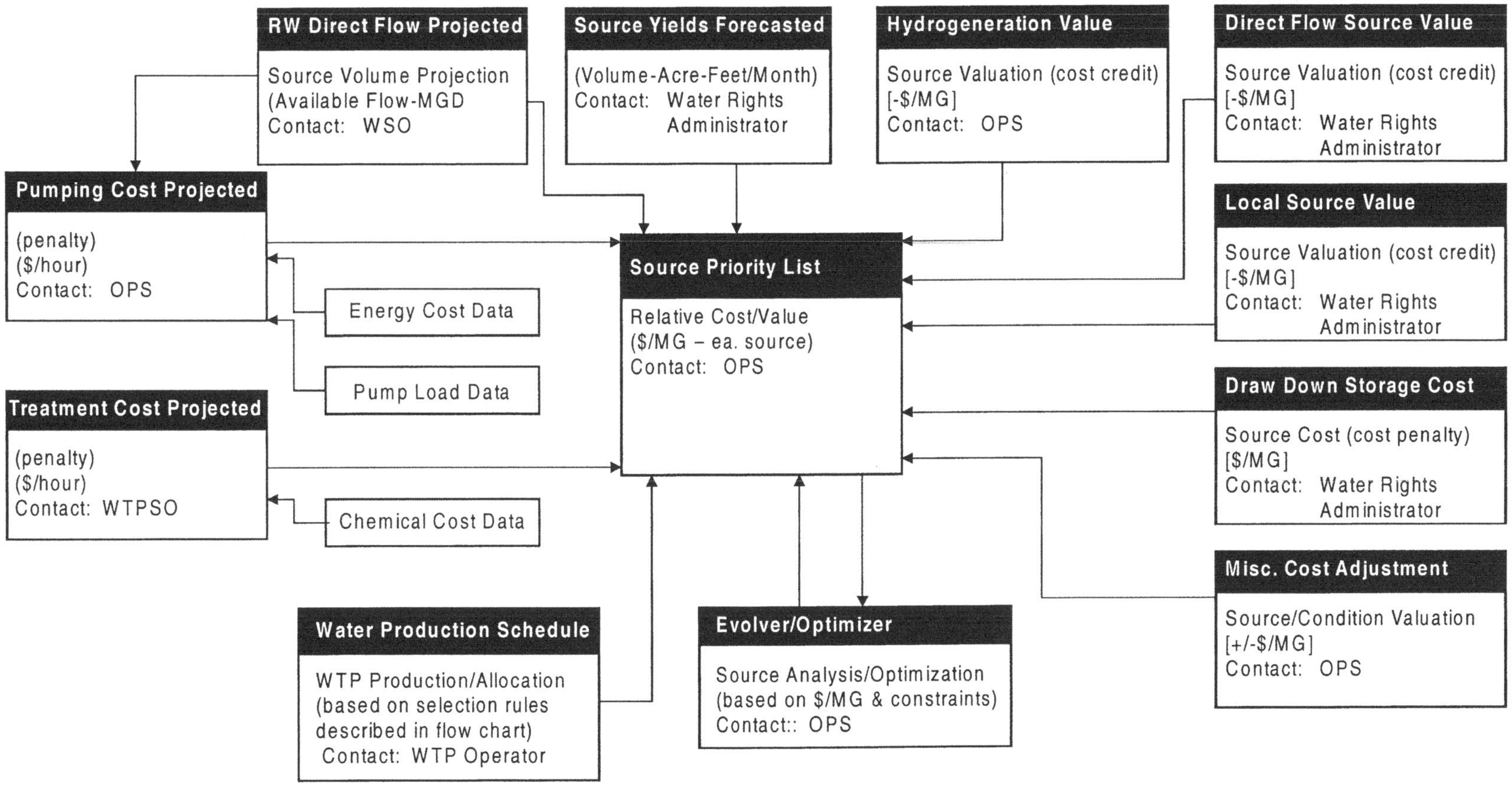

Figure H.1 Water Source Analyzer layout

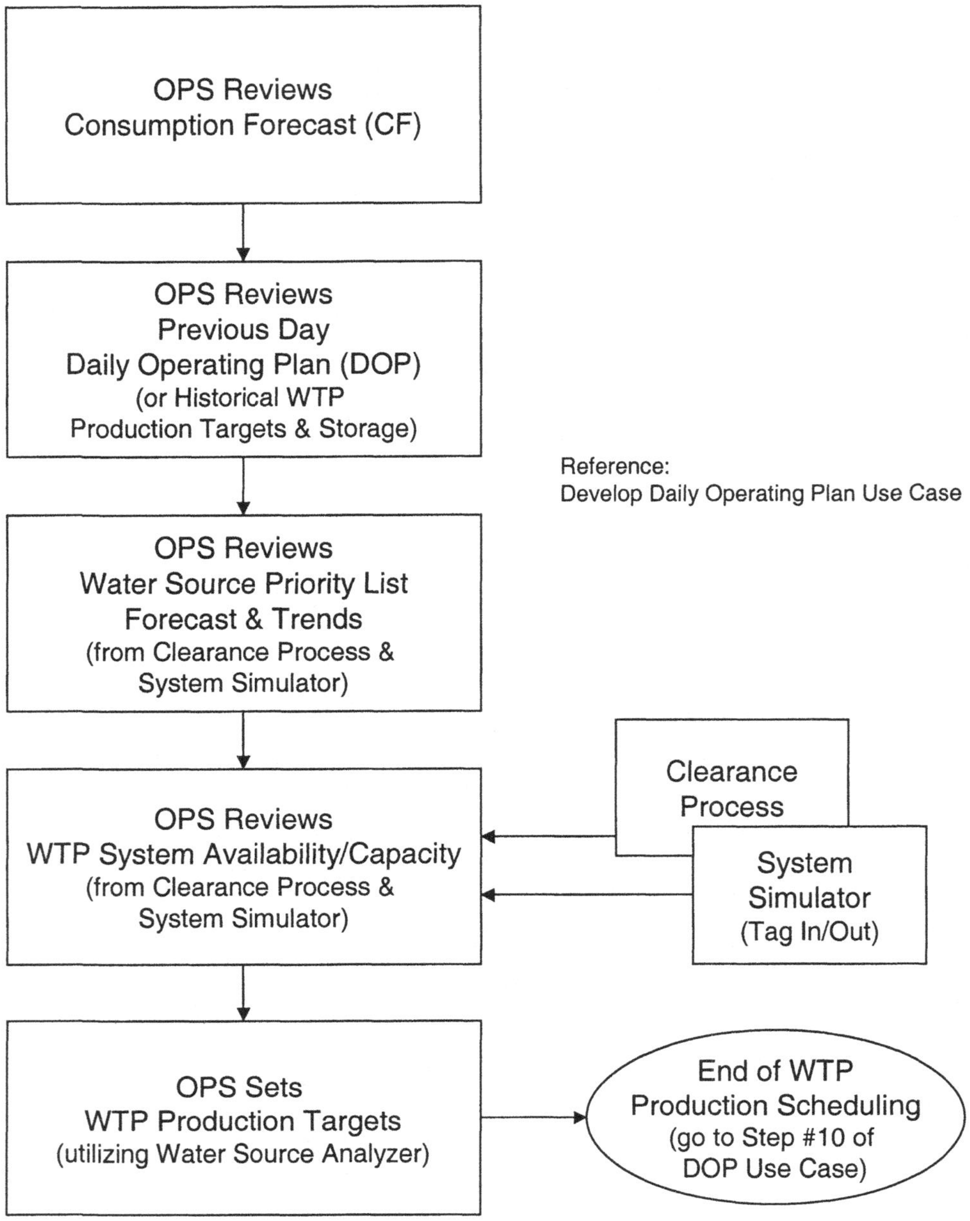

Figure H.2 Water treatment plant production scheduling

APPENDIX I
FUNCTIONAL SPECIFICATION
DAILY OPERATING PLAN (DOP)

PURPOSE

The Daily Operating Plan (DOP) application uses a wide range of information obtained from other EWQMS applications and the SCADA historian databases, to prepare daily water system operating plans. Plans are summarized (graphically) to include large volumes of information including optimized schedules, supporting information (i.e., maintenance activities, weather, energy data) necessary to operate the entire water system.

OBJECTIVE

The objective of the DOP application is to bring together and present optimized schedules provided by the Water System Simulator and supporting information to stakeholders responsible for construction, maintenance, and operation of the water system.

INITIAL IMPLEMENTATION

The initial implementation of the DOP is a spreadsheet application assembled with direct input from SCADA historian databases, transfers from the System Simulator, and manual input of supporting information.

FUTURE IMPLEMENTATION

Future implementation of the DOP application will include provisions for improved presentation capability, capacity and performance, including enhanced data transfer utilizing an integration bus for more robust management of data and communications.

FUNCTIONAL DESCRIPTION

Overview

Once a day, the Operations Planner/Scheduler (OPS) produces a DOP based on data extracted from final optimization run of the System Simulator. Steps for production of the DOP are as follows:

1. Input of assumptions guiding the development of the DOP
2. Input of weather related information (transfer from numerous weather monitoring sources)
3. Input of consumption forecast data including hourly service level and total system consumption profiles
4. Input of construction/maintenance clearance information, including approved and pending clearances (by transfer from the Clearance Procedure application)

5. Input of optimized schedules for pumps, valves, hydro-generators, water source transfers, water treatment plant productions, and tank levels (by transfer from the System Simulator application)
6. Input of energy related information including price, load, value, load shedding (by transfer from the Energy Cost Scheduler application)
7. Development of trend charts for comparison of previous days forecast versus actuals for performance evaluation

Human-Machine Interface (HMI)

Figures I.1 through I.3 illustrate the presentation of the Daily Operating Plan application.

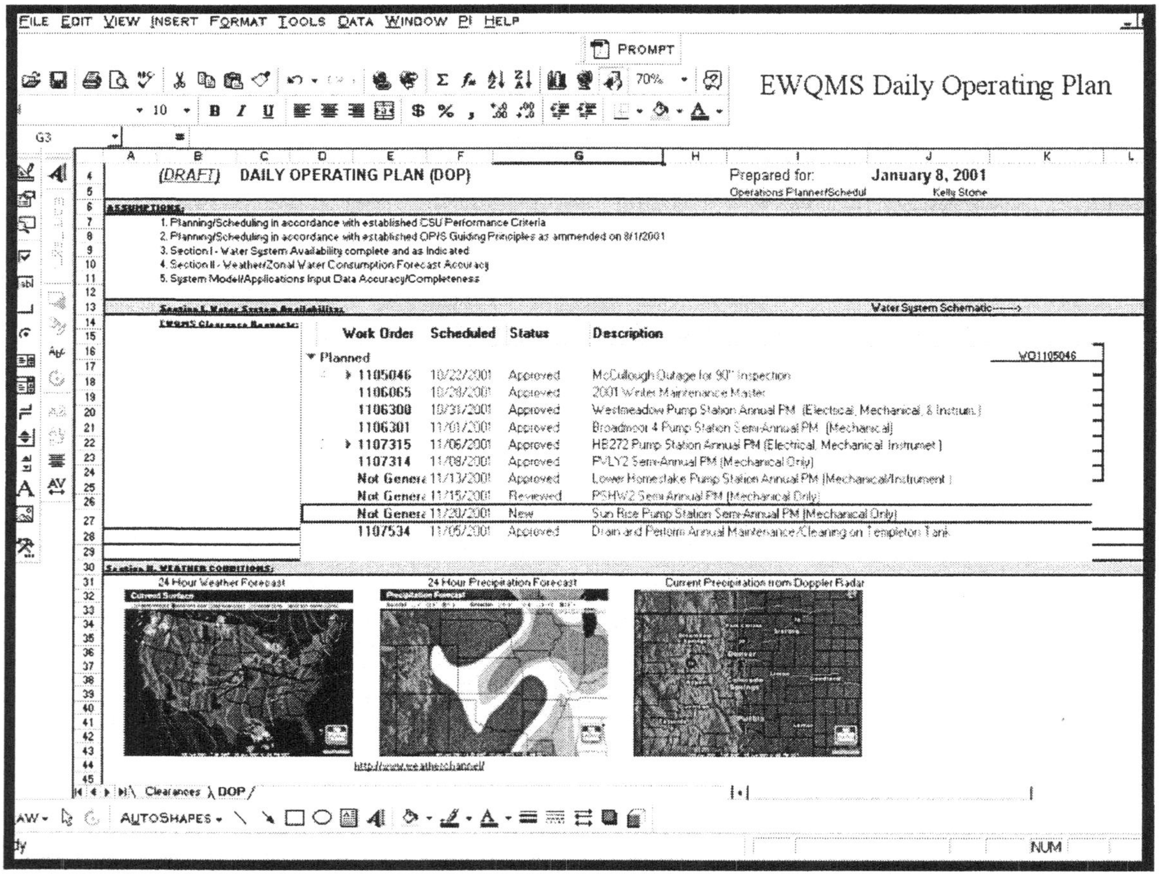

Figure I.1 Daily Operating Plan Page 1 – Assumptions, Clearances, Weather

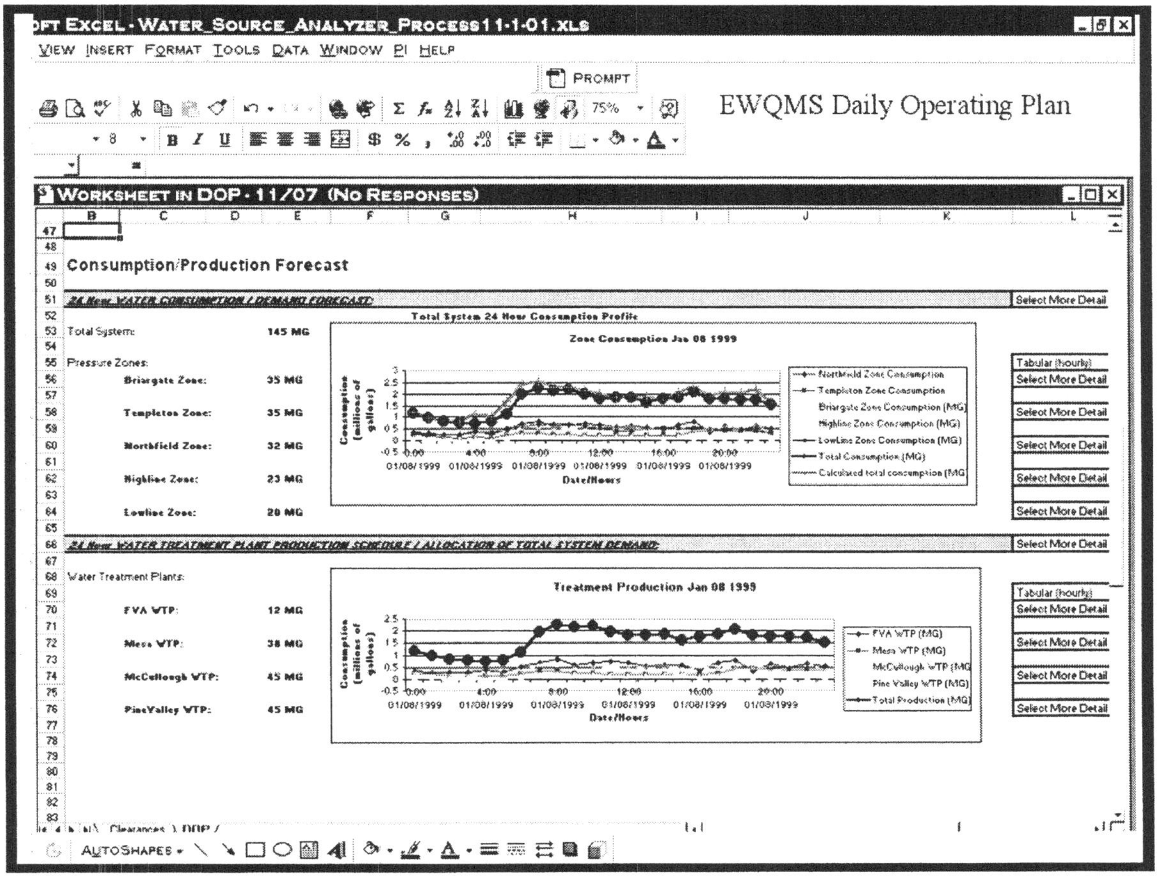

Figure I.2 Daily Operating Plan Page 2 – Consumption Forecast, WTP Production Schedules

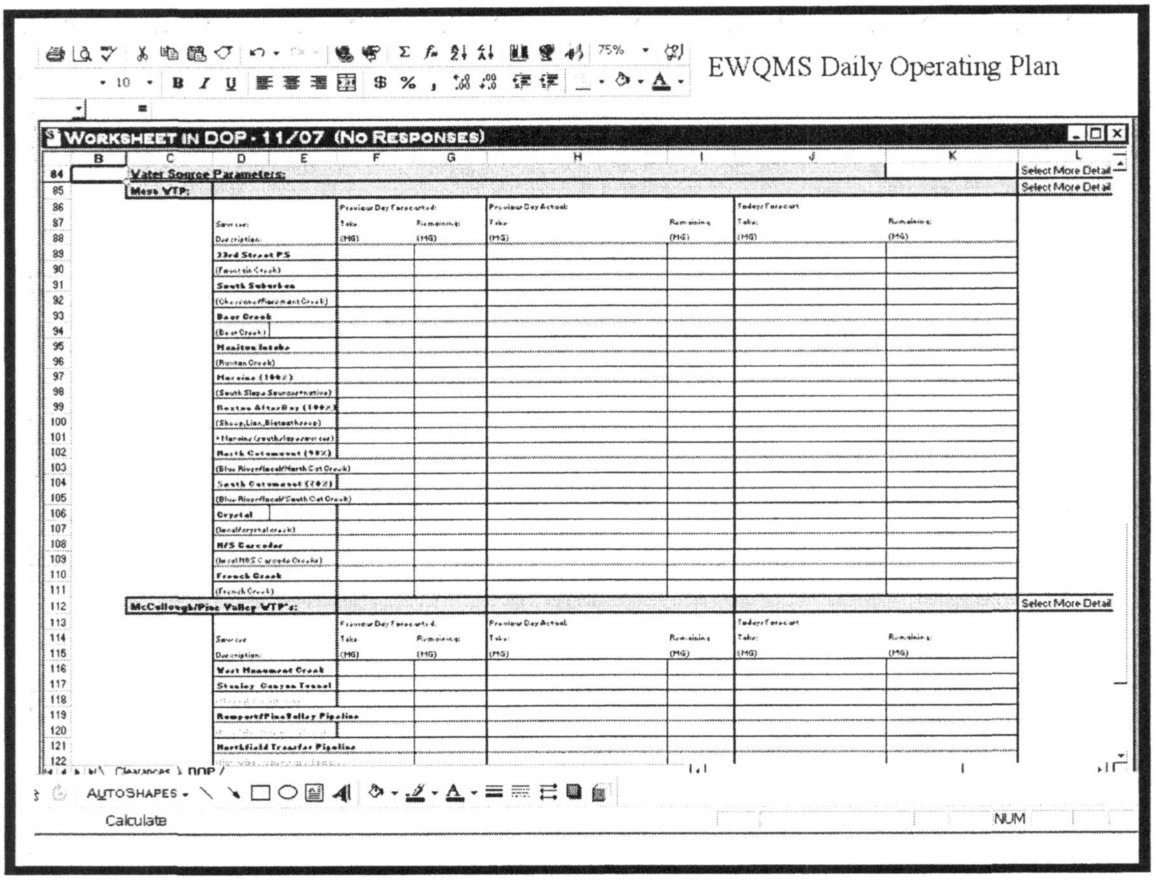

Figure I.3 Daily Operating Plan Page 3 – Water Source Schedule

APPENDIX J
FUNCTIONAL SPECIFICATION
DAILY OPERATING PLAN NETCONFERENCE

PURPOSE

The Daily Operating Plan NetConference application provides a vehicle for rapid and effective sharing of the Daily Operating Plan (DOP). Significant collaboration of the OPS, water system control, treatment, raw water transmission, construction/maintenance, distribution field support, and water quality functions are required to accomplish the significant task of developing the DOP. Plans are summarized (graphically) to include large volumes of information including optimized schedules, and supporting information (i.e., maintenance activities, weather, energy data) necessary to operate the entire water system. Communication of the DOP across a wide area network will meet most needs for sharing data. Further extension of NetConferencing could be accomplished through sharing across the Internet.

OBJECTIVE

The objective of the DOP NetConference application is to facilitate the sharing of optimized schedules and supporting information contained in the DOP with stakeholders responsible for construction, maintenance, and operation of the water system. This information sharing must include opportunity for review, comment, negotiation, rejection, acceptance, approval, finalization, and archival. Additionally, provisions for tracking of these activities, communication about the process, and management of documents are necessary.

INITIAL IMPLEMENTATION

The initial implementation of the DOP NetConference application is a network e-mail discussion database customized to provide the needed functionality. Minimum requirements for the DOP NetConference application include: posting of draft DOPs, finalization of DOPs, editing/approval and archival for DOPs, and tracking of all discussions and modifications to DOPs.

FUTURE IMPLEMENTATION

Future implementation of the DOP NetConference application will include provisions for improved presentation capability, capacity, and performance. Improved integration of DOP spreadsheet documents and change tracking capability will be accomplished. Provisions for greater integration with established EWQMS applications to streamline sharing of data is anticipated.

FUNCTIONAL DESCRIPTION

Overview

Once a day, the Operations Planner/Scheduler submits a DOP based on documents produced by the Daily Operating Plan application. Steps for submittal/use of the DOP NetConference application are as follows:

1. Input of information describing the origin/nature of each DOP document
2. Attachment of the appropriate DOP generated by the DOP application
3. Notification of stakeholders, indicating availability of draft DOP document for review and discussion
4. Discussion of the submitted DOP utilizing discussion thread feature of NetConference application
5. Upon agreement, the DOP is finalized utilizing the approval feature of the NetConference application
6. Finalized DOP is automatically posted in the one day-ahead forecast DOP and is no longer available for edit
7. Archival of final DOPs (and unused drafts) is accomplished by moving past DOPs to the archival folder

Human-Machine Interface (HMI)

The following images illustrate the presentation of the Daily Operating Plan NetConference application.

Figure J.1 displays the one, two, and three day-ahead DOP forecasts folder, the current day-ahead DOP document folder, the current day-ahead draft DOP document folder, instructions for use of the NetConference application, archived DOPs, guiding principles for the DOP, discussion items, and read logs to track utilization of the application. Figure J.2 shows the attached DOP document, response function to comment on the DOP, and the update log to track reviews of the DOP. Figure J.3 displays the response function, allowing categorized response and discussion thread for each DOP document. Figure J.4 shows the current DOP folder. Figure J.5 displays the final DOP. Figure J.6 shows the one-day ahead DOP forecast folder. Figure J.7 shows the archived DOP folder, and Figure J.8 displays the instructions folder for using the DOP NetConference application, the OPS guidelines documentation folder, the documents folder, and the read logs folder. Figure J.9 displays the Read Logs for tracking access/modifications to the DOP.

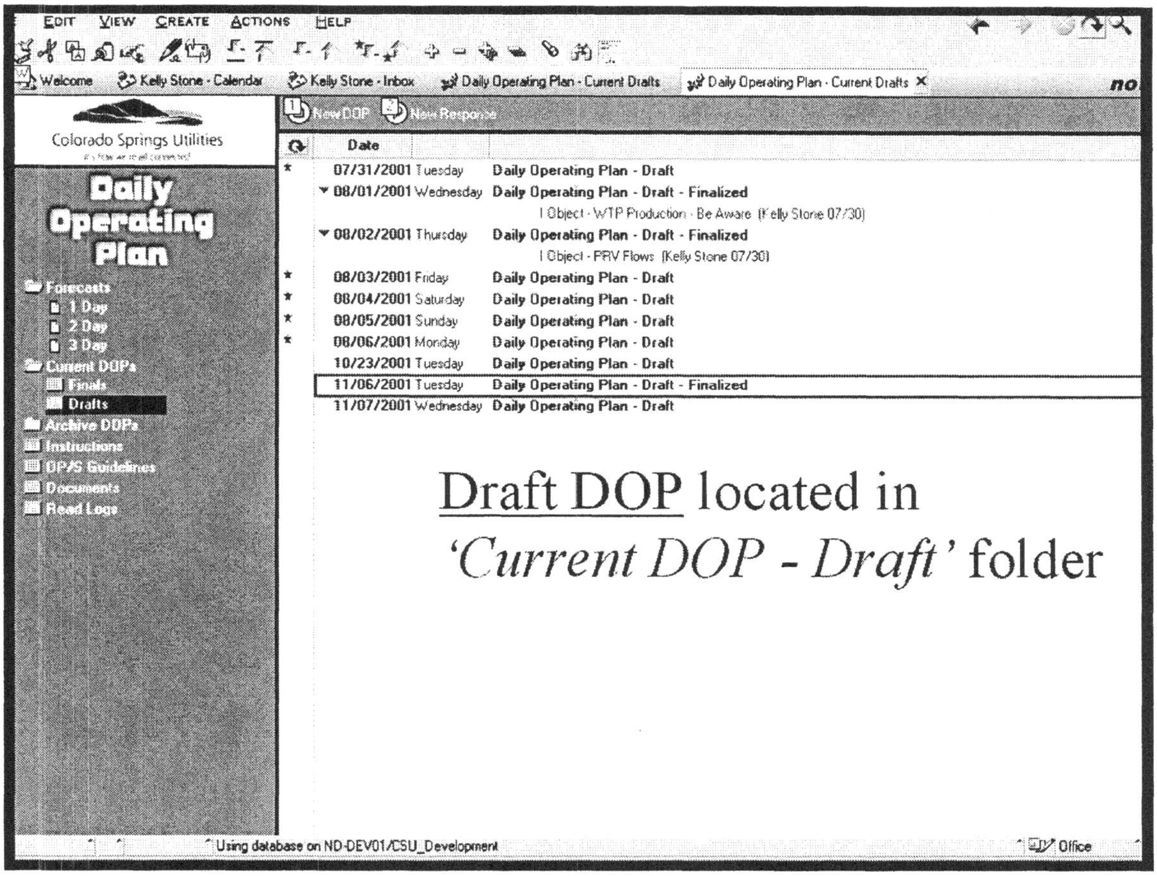

Figure J.1 DOP NetConference Page 1

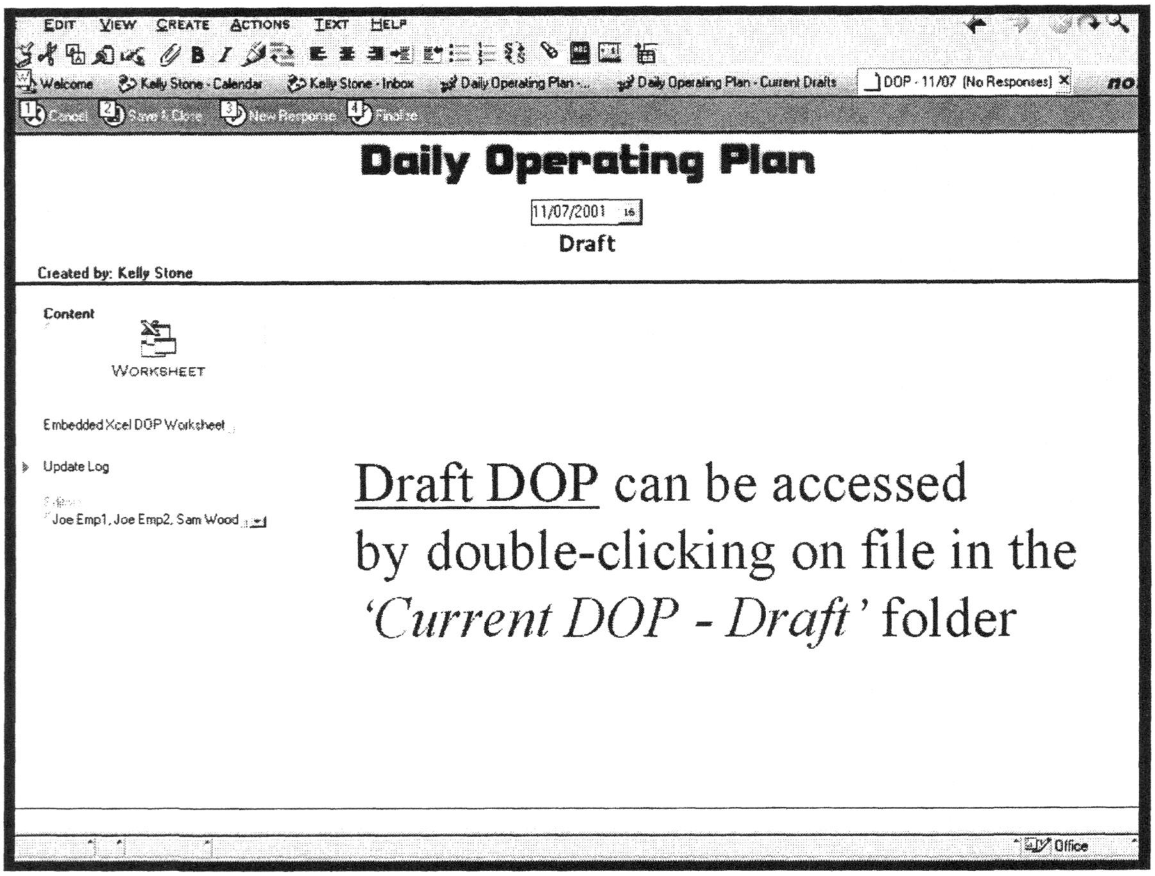

Figure J.2 DOP NetConference – Draft DOP Display

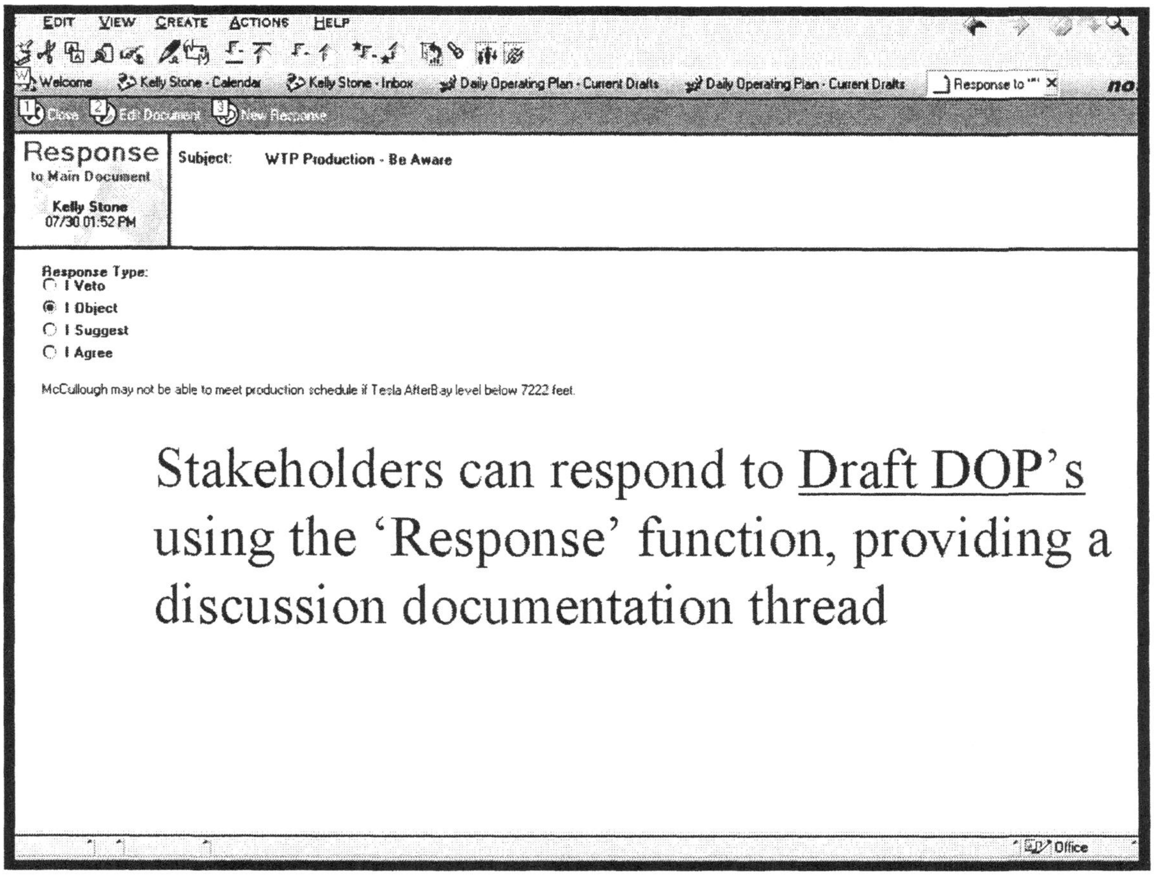

Figure J.3 DOP NetConference – Response Function

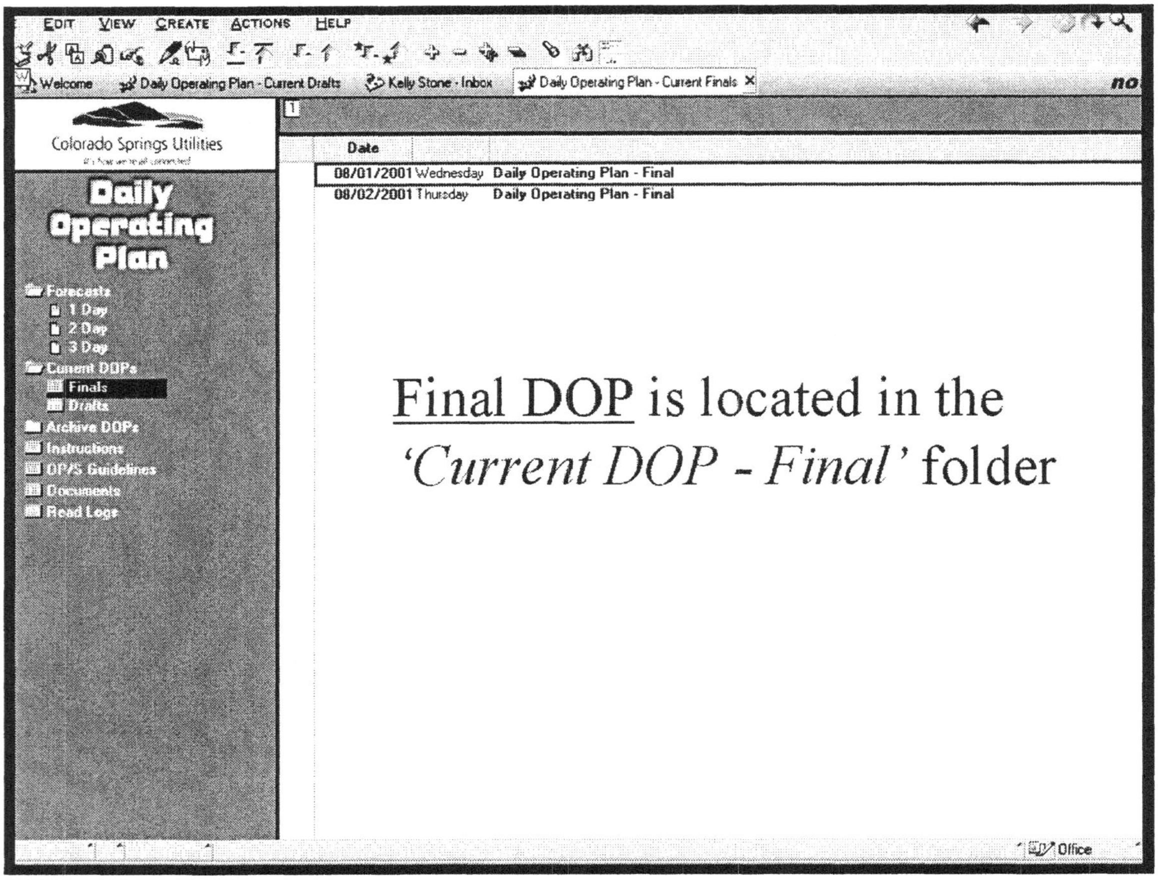

Figure J.4 DOP NetConference – Current-Final DOP Folder

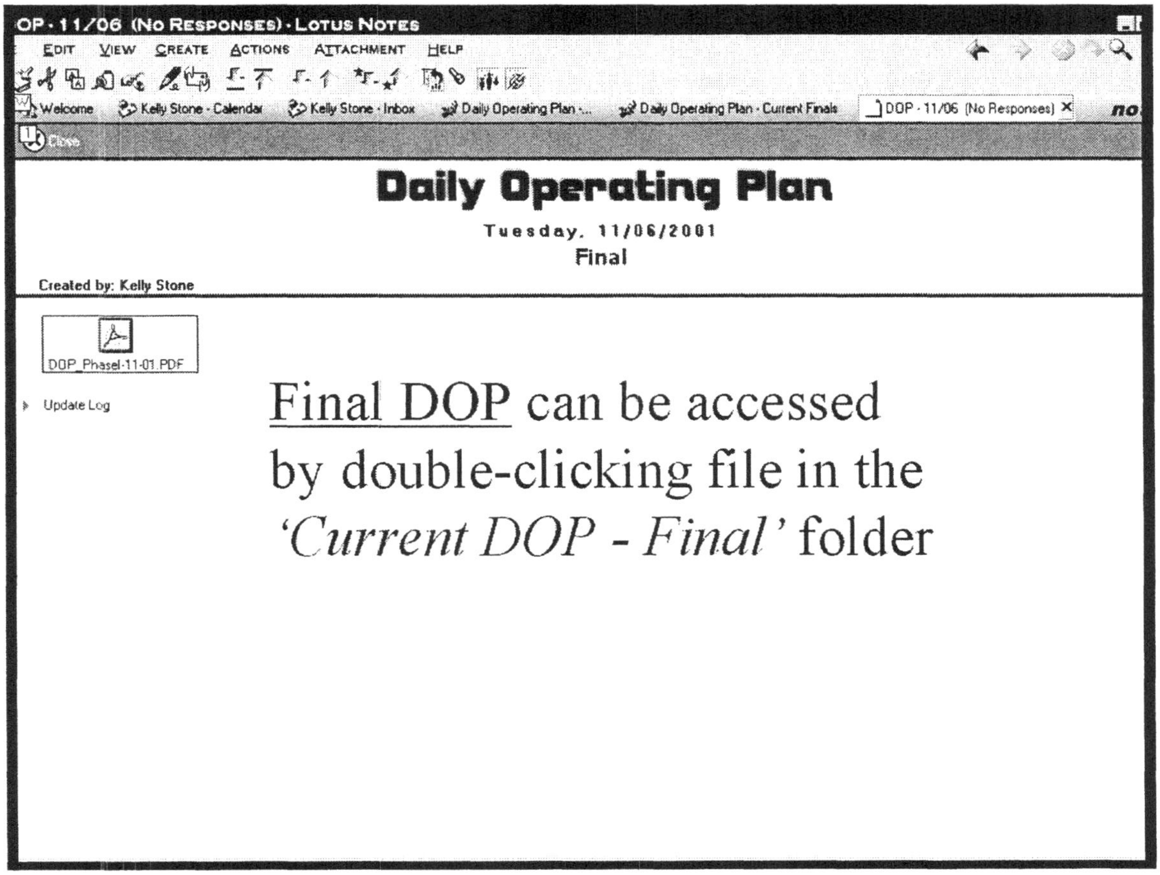

Figure J.5 DOP NetConference – Current-Final DOP

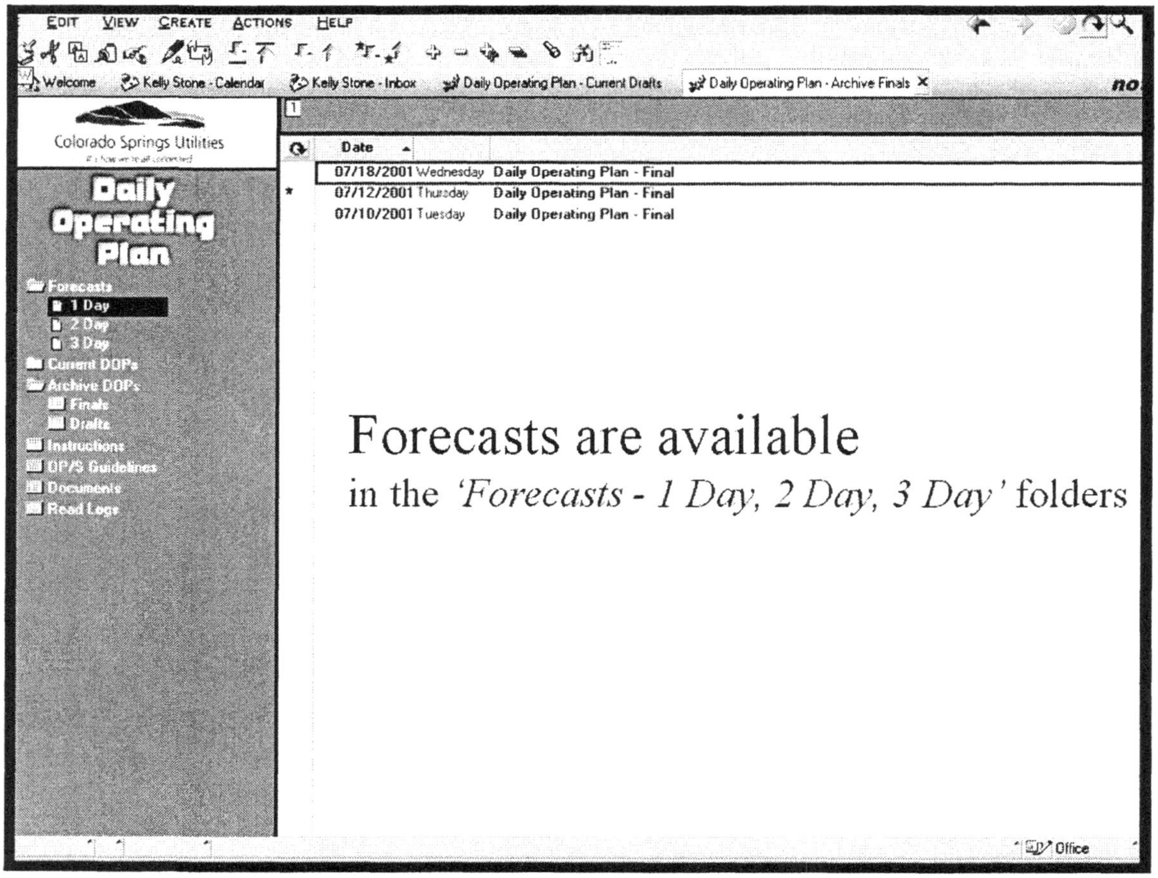

Figure J.6 DOP NetConference – Day-ahead DOP Forecast Folder

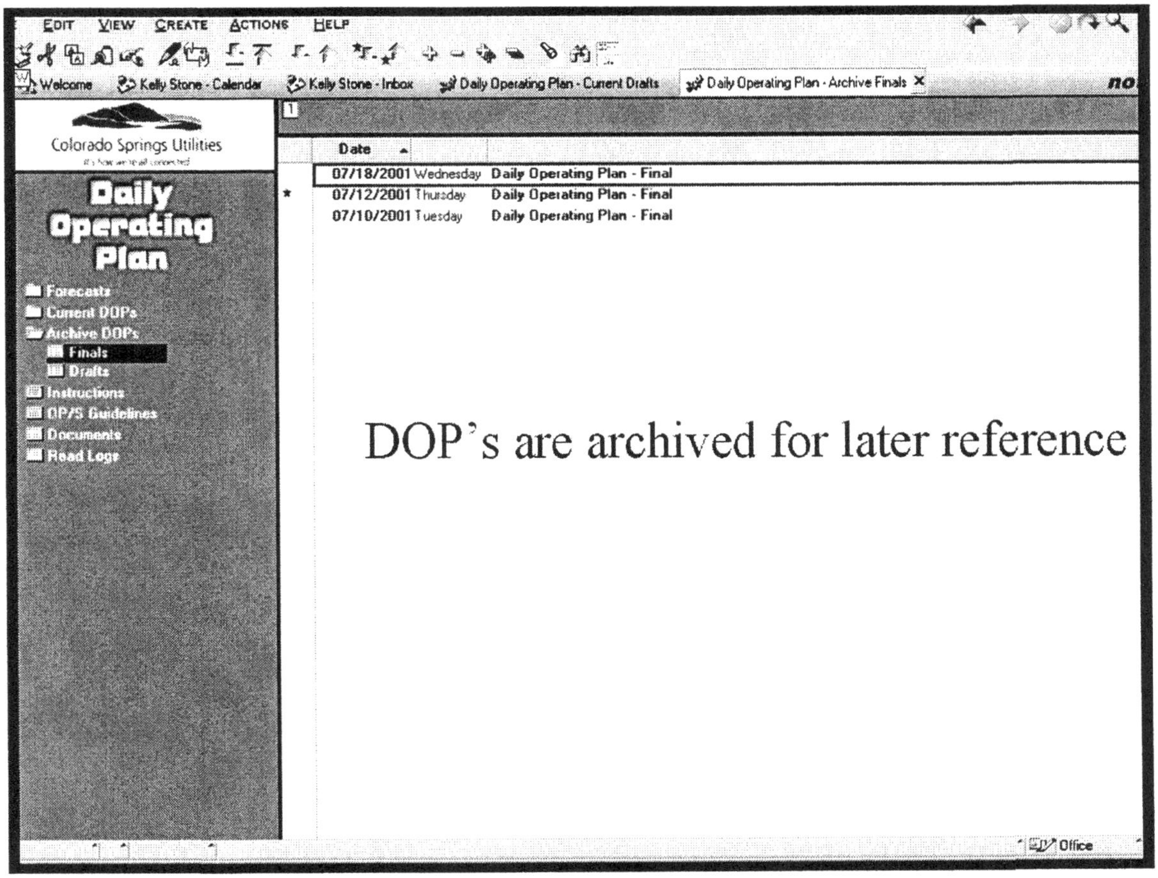

Figure J.7 DOP NetConference – Archive DOP Folder

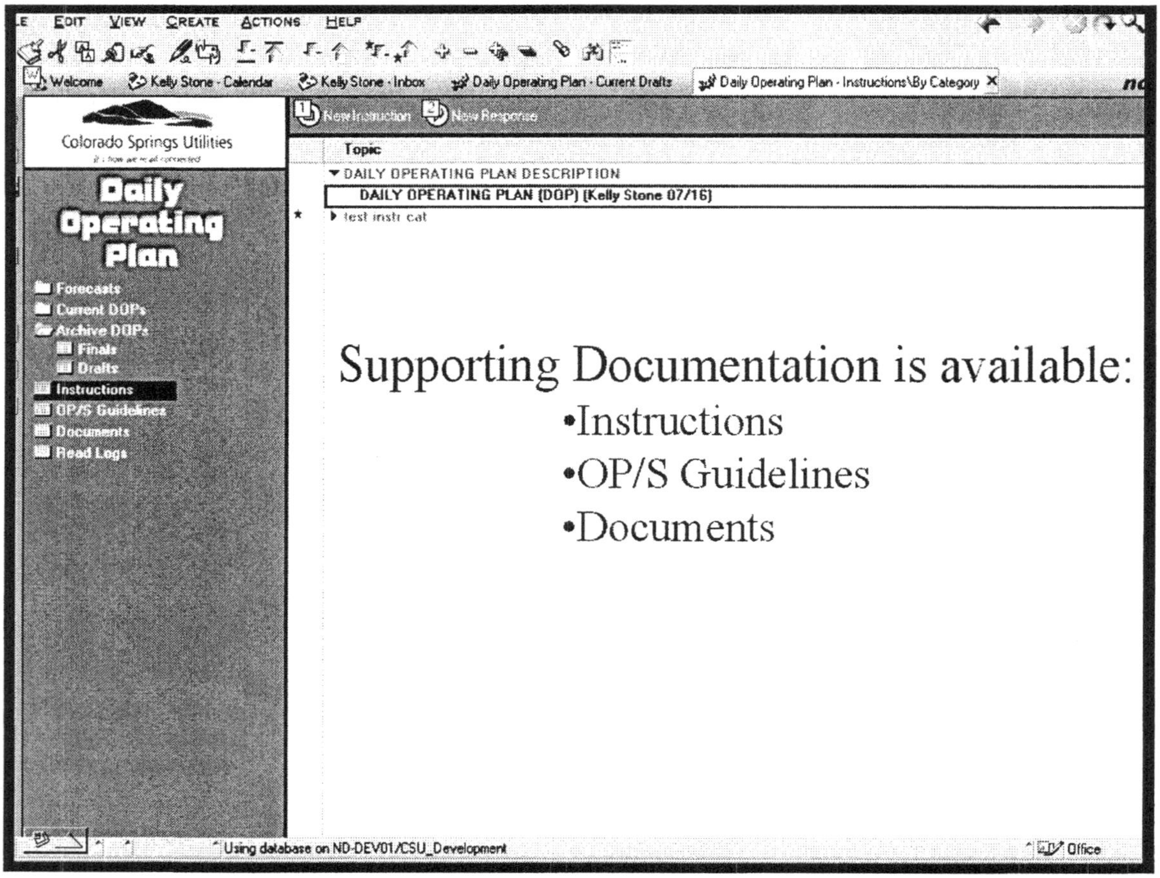

Figure J.8 DOP NetConference – Instructions, OPS Guidelines, Documents, Read Logs

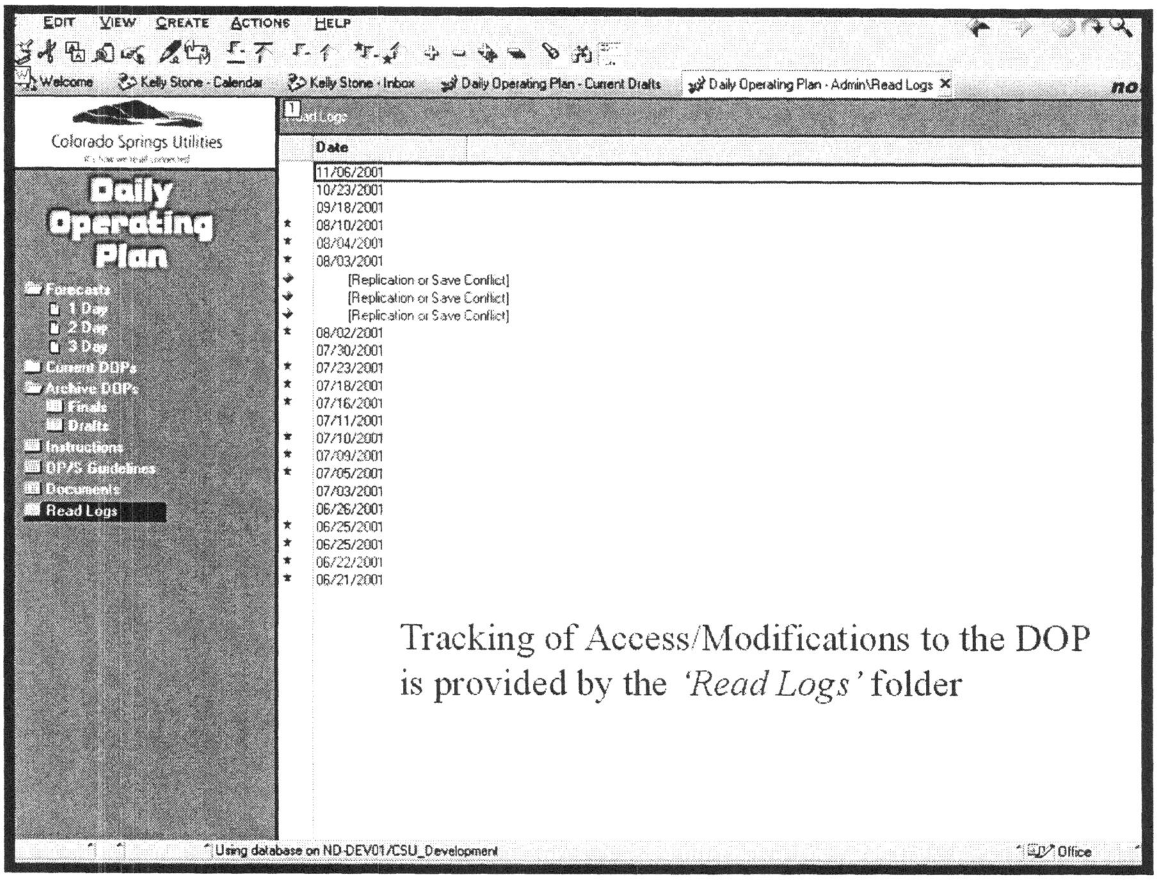

Figure J.9 DOP NetConference – Read Logs Folder

APPENDIX K
FUNCTIONAL SPECIFICATION
CONSTRUCTION/MAINTENANCE CLEARANCE PROCESS

PURPOSE

The purpose of the Construction/Maintenance Clearance Process (CP) is to coordinate operations of water system components that affect system operations reliability and safety of the CSU work force.

OBJECTIVES

Objectives of the Construction/Maintenance CP are the following:

- Provide the water system work force with the information they need to plan and schedule in order to implement the Clearance Process
- Provide employees with the information they need to identify their areas of responsibility and perform their job duties
- Ensure adequate system availability to support the execution of the Daily Operating Plan (DOP) through planning and scheduling water system construction and maintenance activities, which impact water system operations
- Provide customers a quality product at a reduced cost while minimizing system disruption, improving safety for CSU crews and maintaining security of the water system

INITIAL IMPLEMENTATION

The initial implementation of the EWQMS Clearance Process application will require the use of a discussion group function of an email application that has been customized for the purpose of tracking and documenting construction/maintenance activities. Access lists to restrict privileges for the authoring, approval and administration of clearances will be utilized. Documents status will be assigned and each document archived upon conclusion of active clearance usage. The OPS will provide administration of the CP application. Approval of CP requests will be provided by the OPS function including the Water System Operators (WSOs).

FUTURE IMPLEMENTATION

Future modifications of the EWQMS Clearance Process application may include the transition to use of a more general-purpose work management system allowing increased integration of clearance data and avoiding duplicate data entry activities.

Additional scope and applicability may be necessary to accommodate future system expansions and organizational structure changes.

FUNCTIONAL DESCRIPTION

A Clearance Request will be required on any activity that would require the WSO to make an adjustment to the Water system. The procedures listed below are to be followed when requesting approval for planned activities.

Preliminary Data Analysis

A four-week pilot of the EWQMS Clearance Process is required to demonstrate CP functionality. Changes in business processes and improved completeness/accuracy of organizational usage are needed for OPS planning and scheduling and DOP preparation.

Planning the Work Activity

The Construction/Maintenance Scheduler (Planner) of the work group planning the activity identifies, evaluates, estimates and schedules the planned activities in advance.

1. Identifies - Through communication and collaboration with organizational entities, activities are identified.
2. Evaluates - The work activity is evaluated as to the impact on the water system. If upon evaluation it is determined the activity will have a major impact, then the Clearance Request Process will be required, and the Construction/Maintenance Scheduler (Planner) will prioritize the work activities.
3. Estimates - For each work activity, the labor, materials, and tool requirements will be estimated along with target start and completion dates and times. If an outage for a customer(s) will occur, the estimated duration will be calculated.
4. Scheduling - Each work activity will be scheduled as far in advance as possible and updated as better information becomes available.
5. Communications - The Scheduler (Planner) will complete the Clearance Request Form for the "Planned" section as found in the EWQMS CP application. The form will be forwarded to the OPS (or WSO acting as the OPS) for approval.

Using the Clearance Request Data

The OPS and/or WSO will review the request, approve it, or will contact the Construction/Maintenance Scheduler (Planner) or Site Contact to resolve any conflicts which might affect the date and time the activity is scheduled. The OPS/WSO will base their approval decision on a first come first serve basis unless an unscheduled event takes priority. Further communication and collaboration will be initiated by the OPS/WSO if required.

If the Clearance Request is rejected by the OPS (or WSO acting as the OPS) they must communicate with the Construction/Maintenance Scheduler (Planner) or Site Contact to resolve issues to approve the request. This may result in rescheduling conflicting activities.

Site Contact Responsibilities

Prior to beginning an activity, it is the Site Contact's responsibility to furnish the OPS/WSO with updated data regarding the Clearance Request. It is also the Site Contact's responsibility to communicate with the WSO (via radio or phone call) when this activity is to start. Once the activity has been completed, the Site Contact will also communicate with the WSO when this activity is completed.

Actuals/Unscheduled Events

When an emergency or unscheduled event occurs, the Site Contact communicates with the WSO (by radio or phone) the actual information, so the WSO can make adjustments to the water system. When the unscheduled activity is completed, the Site Contact communicates (by radio or phone) with the WSO so the WSO can make adjustment to return the system to normal operating settings and document those adjustments.

Human-Machine Interface (HMI)

The following figures describe the presentation of the EWQMS Clearance Process application:

Figure K.1 illustrates the Clearance Request General Information Display that includes: status, outage duration, site contact information, comments, work order number, activity description, vicinity, and dates for planned, scheduled, and actual performance of activity. Figure K.2 displays the Clearance Request Planned Equipment Log in which all planned equipment status changes are noted with planned date/time of status change. Figure K.3 illustrates the Clearance Request Actual Equipment Log in which all planned equipment status changes are noted with actual date/time of status change. Figure K.4 displays the Clearance Request Planned and Unplanned Folders that include the planned clearance document folder and the unplanned clearance document folder. Note that all folders and individual clearance headings can be sorted by the categories in the left column. Figure K.5 displays planned and unplanned clearances sorted by work order. Attached responses to individual Clearance Requests are shown.

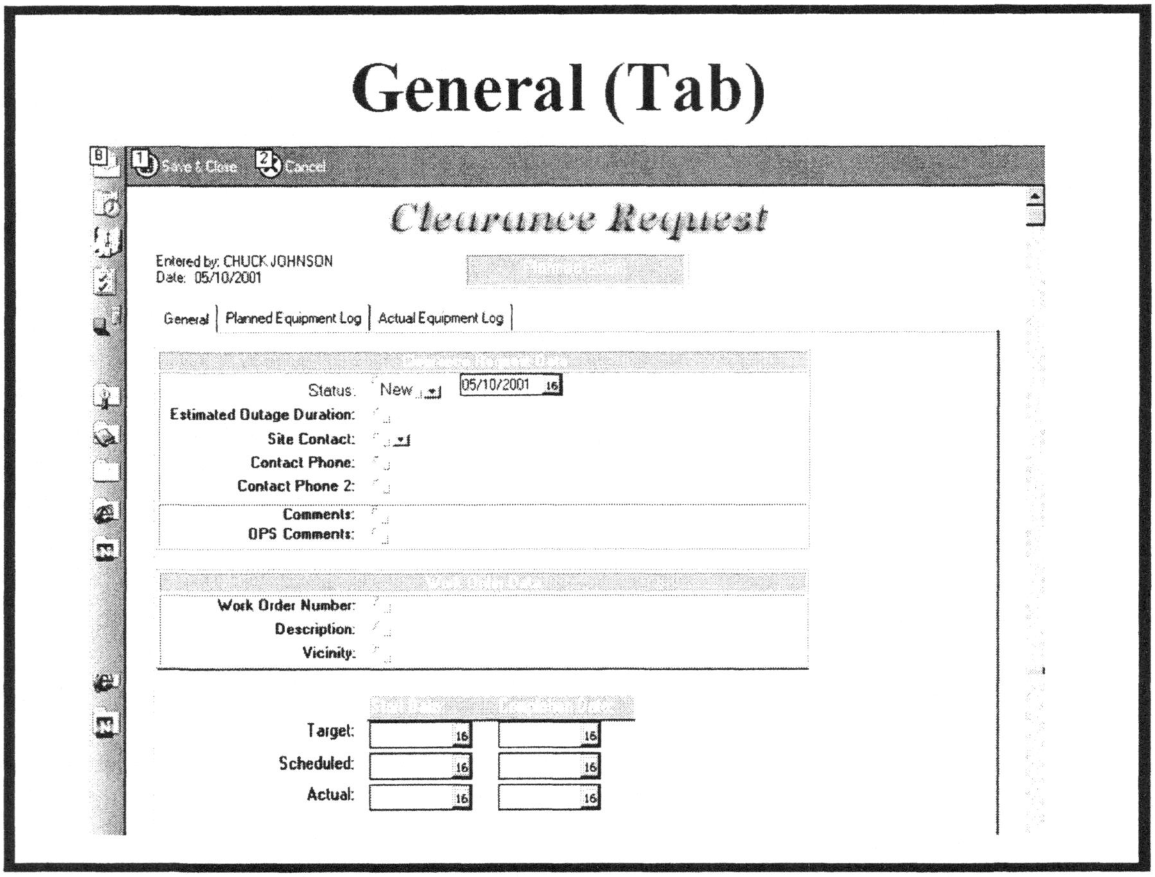

Figure K.1 Clearance Request – General Information

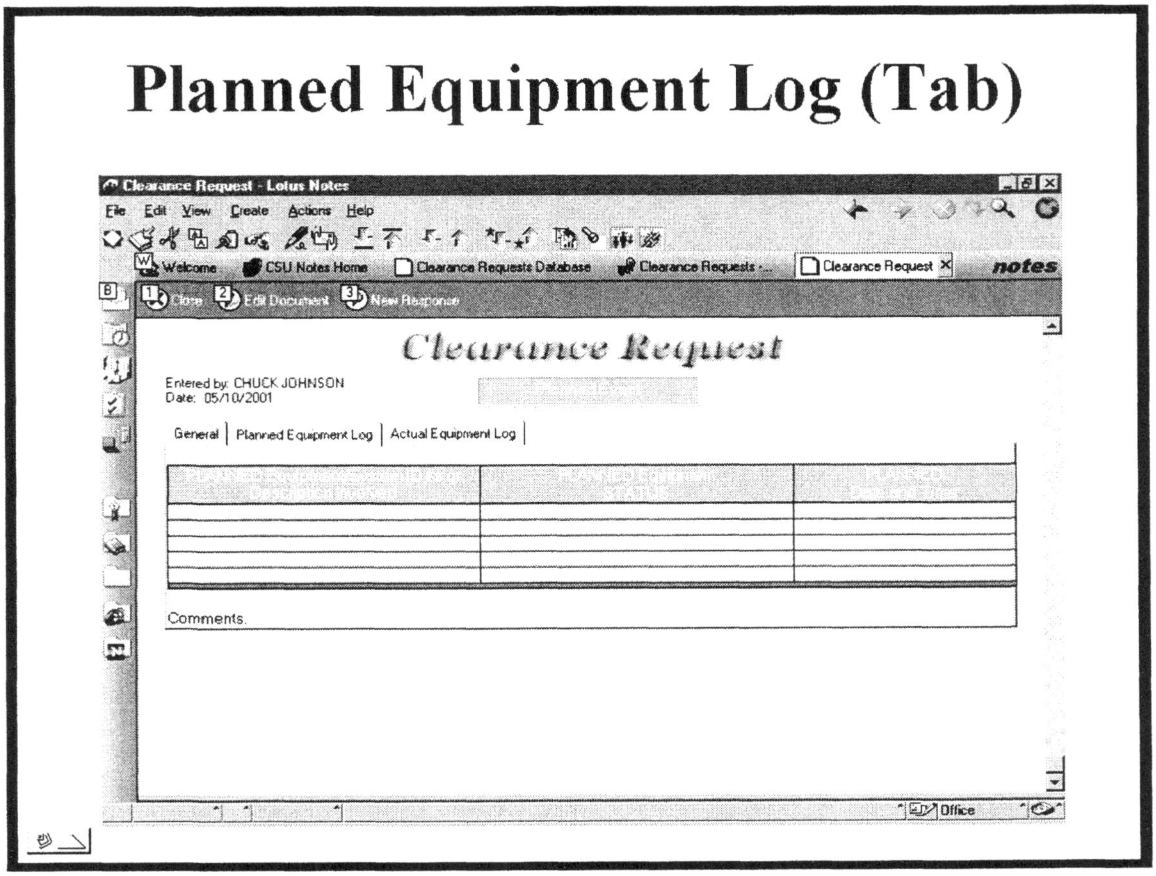

Figure K.2 Clearance Request – Planned Equipment Log

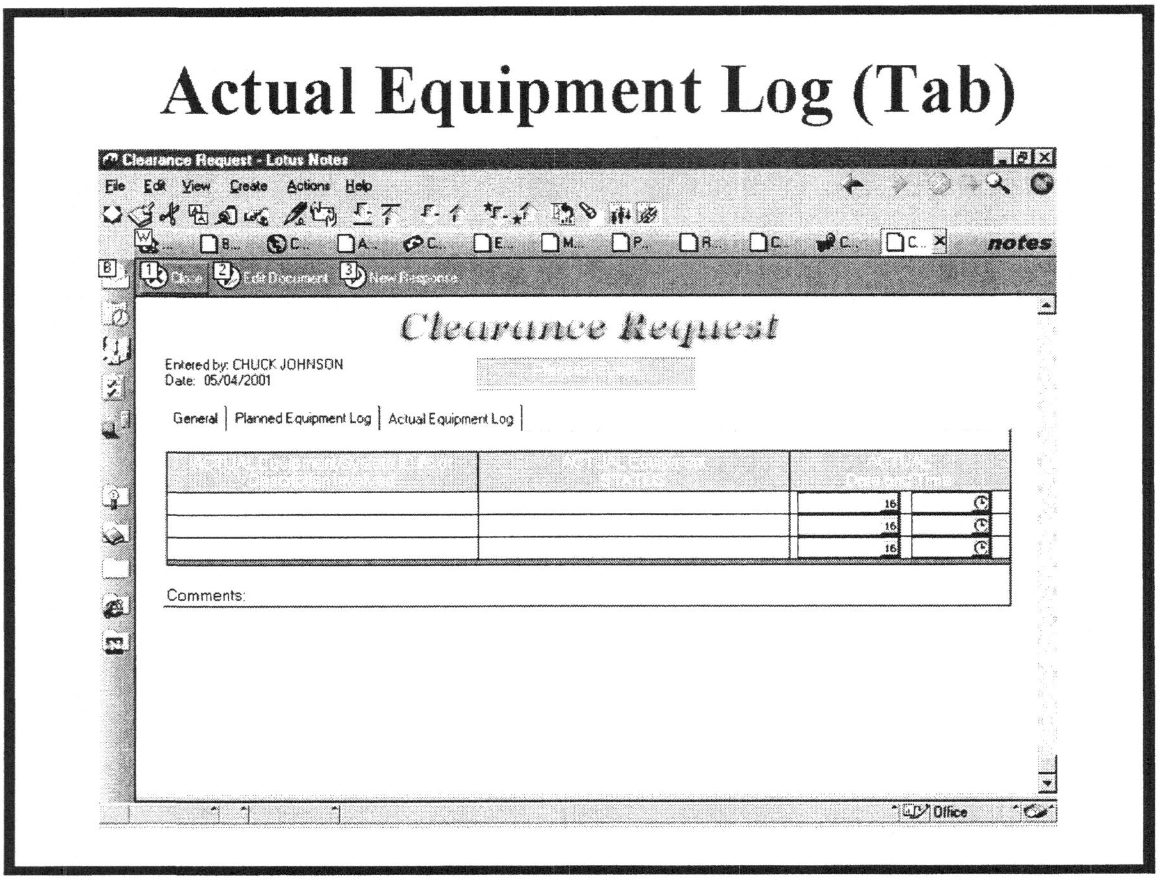

Figure K.3 Clearance Request – Actual Equipment Log

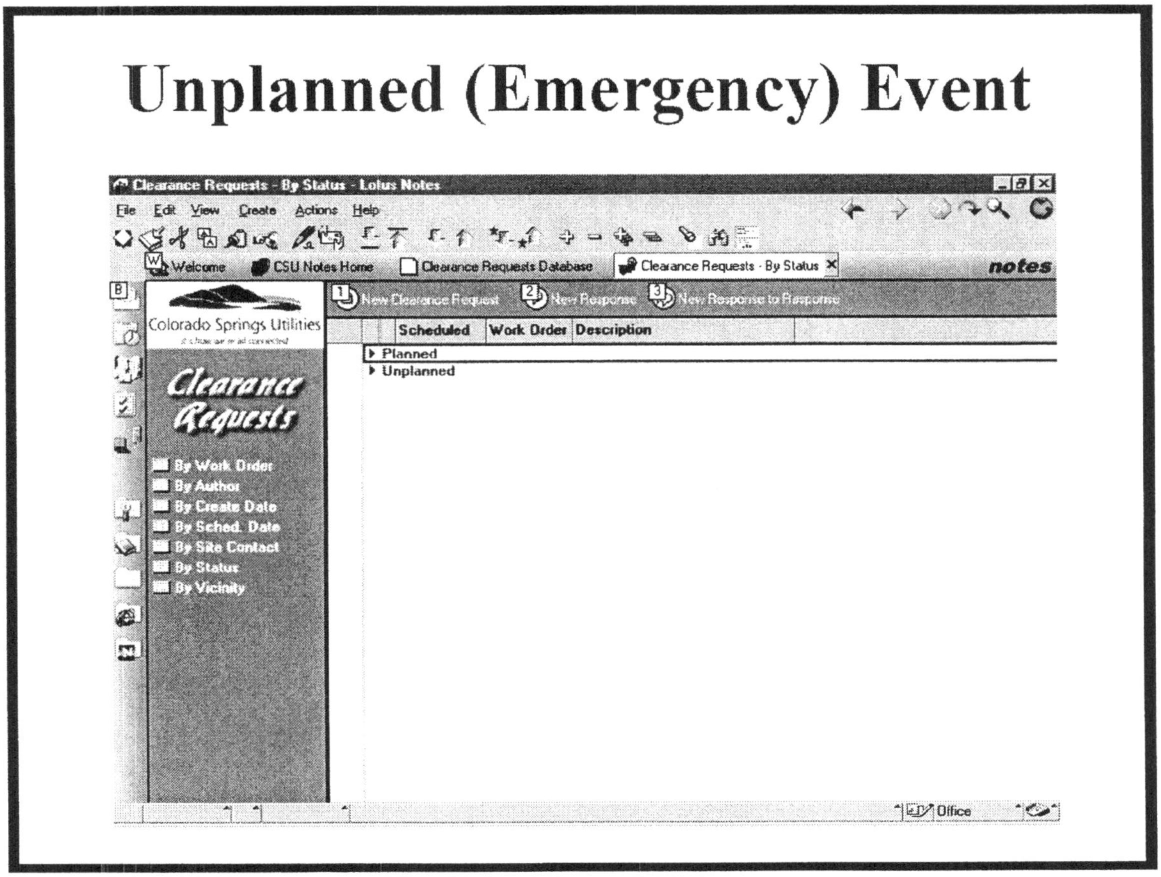

Figure K.4 Clearance Request – Planned and Unplanned Folders

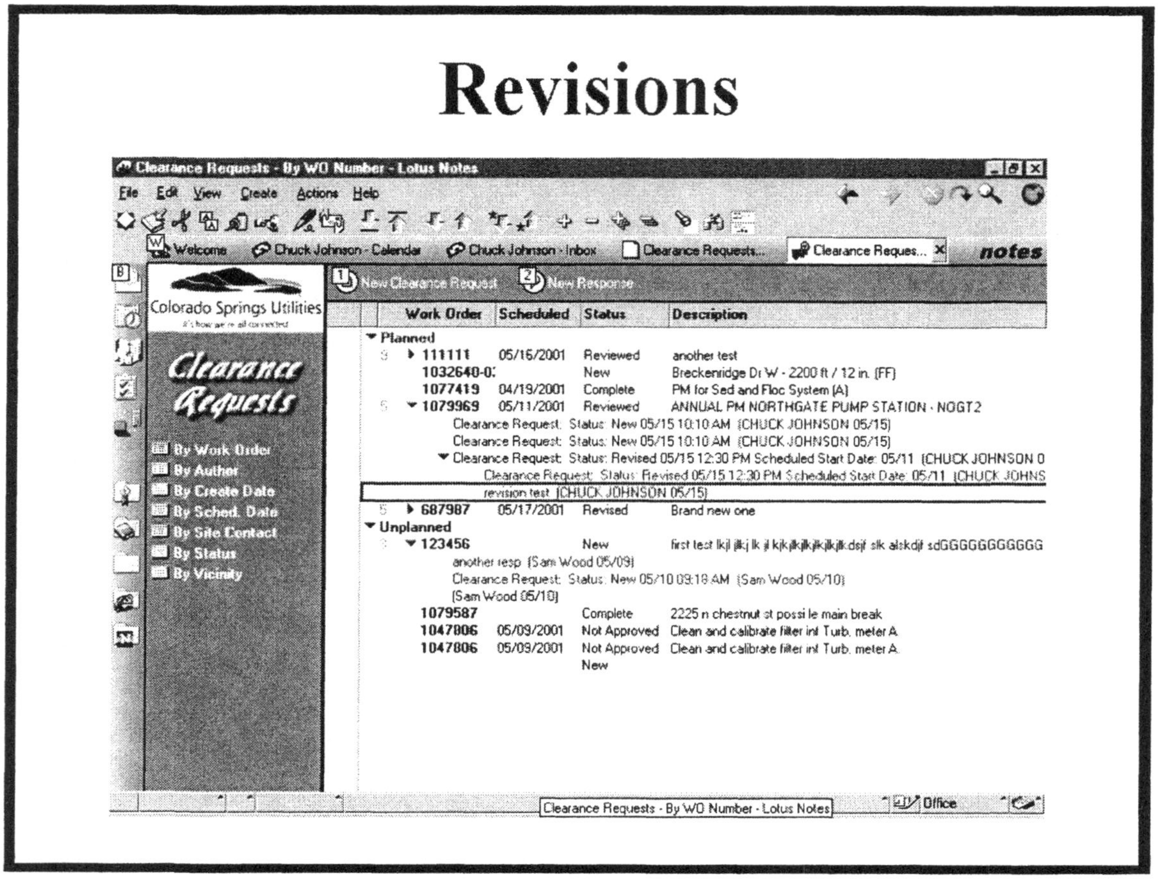

Figure K.5 Planned and Unplanned Clearance Requests

REFERENCES

Brueck, T., M. Rettie, M. Rousso. 1998. *The Utility Business Architecture: Designing for Change.* Denver, Colo.: Awwa Research Foundation (AwwaRF) and American Water Works Association (AWWA).

Grayman, W., L.A. Rossmann, C. Arnold, R.A. Deininger, C. Smith, J.F. Smith, R. Schnipke. 2000. *Water Quality Modeling of Distribution System Storage Facilities.* Denver, Colo.: AwwaRF and AWWA.

Jentgen, L., D. Curtice, J. Jacobs, E. Kawczynski, K. Carns. 1997. *Energy and Water Quality Management System: Collaborative Project for Water System Optimization.* St. Louis, Mo.: EPRI (Electric Power Research Institute) Community Environmental Center.

ABBREVIATIONS

AI SCADA	Advanced Integrated Supervisory Control and Data Acquisition
AwwaRF	Awwa Research Foundation
CP	Clearance Procedure
CSU	Colorado Springs Utilities
DOP	Daily Operating Plan
Econ1, Econ2	Economic Control Mode 1, Economic Control Mode 2
EPA	Environmental Protection Agency
EPRI	Electric Power Research Institute
EPS	Extended Period Simulation
EPVC	Economic Pump/Valve Controller
ET	Evapo-Transpiration
EWQMS	Energy and Water Quality Management System
(F)	flow
FCV	Flow Control Valve
GIS	Geographic Information System
gpm	gallons per minute
GUI	Graphical User Interface
HMI	Human-Machine Interface
HOA	Hand/Off/Auto
I/O	Inputs/Outputs
I-Team	Implementation Team
kgals	Kilo Gallons
kwh	Kilowatt Hour
LAN	Local Area Network
LIMS	Laboratory Information Management System
L/R	Local/Remote
MGD	Million Gallons per Day
mg/L	milligrams/Liter
min	Minutes
ODBC	Open Database Connectivity
ODC	Organizational Development Consultant
OPS	Operations Planner and Scheduler

OR	Logical OR Function
(P)	Pressure
PCS	Plant Control System
PRV	Pressure Reducing Valve
PSI	Pounds per Square Inch
ROI	Return-On-Investment
RTU	Remote Terminal Unit
SCADA	Supervisory Control and Data Acquisition
SLT	Strategic Leadership Team
SME	Subject Matter Expert
SQL	Structured Query Language
TOU	Time-Of-Use
UBA	Utility Business Architecture
WRD	Water Resources Department
WSA	Water Source Analyzer
WSO	Water System Operator
WTP	Water Treatment Plant

IWA Publishing's authorised EU representative for General Product Safety Regulations is Diane D'Arras, 15 rue Duret, 75116 Paris, France, e-mail: safety@iwap.co.uk.

Printed and bound by CPI Group (UK) Ltd, Croydon, CR0 4YY
17/09/2026
02218813-0001